JUSTIN GLASER

Sweat: Uncovering Your Body's Hidden Superpower

Contents

Praise for Sweat

"An essential read for anyone seeking a deeper understanding of health, detoxification, and the transformative power of sweat." **Katie Kaps, co-founder of Higher Dose**

"I didn't go into *Sweat* expecting it to be an eye opener, but it was. Strongly recommend to anyone looking to level up their health." **Geri Topfer, Founder of Kula for Karma**

"A truly fascinating journey into a remarkably neglected aspect of our well being. *Sweat* is a must-read for anyone committed to optimal health in the modern world." **Dan Root, President of PureBeing**

"While everyone has some inkling of the importance of nutrition, there is little talk about why sweating is so important. Justin does a great job bringing that to the forefront in a concise and easily understandable manner. " **Kyle Kinkead, Founder of NCP**

"The information in this book has been needed for a long time. As a 9/11 survivor exposed to a cataclysmic toxic event, my body was shutting down. I used equipment to breathe. No doctor knew how to help me. I found a Detox Project using the Niacin Sauna Protocol that was treating sick responders and survivors.

Within three months, the visible physical and mental health turnaround was incredible. Twenty-two years later, 6,000 more people have passed from that exposure, yet almost all of the people who did this protocol are still here. I believe it saved and extended my life. *Sweat* does a terrific job of breaking down some of the most important healing information on the planet."**Anne-Marie Principe, 9/11 survivor and activist**

I

Sweat

"It's not that we don't know so much. It's that we know so much that isn't so." – Paul Dirac

1

Introduction

Imagine you're walking through a bustling city on a hot summer day. The sun is beating down, and you can feel the sweat trickling down your back. It's uncomfortable, it's inconvenient, and you can't wait to get back to the comfort of your air-conditioned home. But what if I told you that sweat, this seemingly insignificant byproduct of a hot day, could hold the key to a health revolution?

Stumbling into Sweat: My Story

Health was never on my mind until I turned 20. Upon graduating from college in 2017, I planned to do a once-in-a-lifetime hike with one of my closest friends. We mapped out the Annapurna Circuit, an epic walking trail through a stunning landscape high in the Nepalese mountains. This was no Ironman excursion, but it did require significant training and preparation.

Upon returning home from school, it was evident to my family that something was amiss. My parents, genuinely concerned, expressed doubts about my upcoming trip to Nepal. I knew deep down they were right. I looked unwell. Through pangs of shame and guilt and with a large lump in my throat, I had to call my friend and let him know I'd be canceling at the last minute. As much as it pained me, I knew I wouldn't be able to successfully make it through the hike. There was something wrong with my health, and hiking halfway across the world wasn't going to work.

By 2017, I was experiencing chronic fatigue, depressive moods, brain fog, and massive anxiety. Though I was skinny and athletic, my body felt like it always needed rest. This had developed so slowly and without drama that I barely questioned it. I had grown used to feeling this way, as it hadn't happened all at once. Everything in life was experienced through this mode of being. It was like forgetting I was wearing a pair of dark sunglasses, tinting my entire reality. Without a reference point for how much better I could feel, it was hard to find motivation or reasons to improve my situation. When I moved from college to NYC, I finally began searching for ways to heal.

My search began in the usual places. I scheduled appointments with Western doctors, hoping they would hold the key to unlock the mystery of my symptoms. "Clean up your diet," some advised. "Get better sleep," others tried. "Let's do bloodwork to see if there's anything in your genetics," one prescribed. I diligently followed their instructions, placing my trust in their expertise. Yet, despite my unwavering commitment, my health problems persisted.

Frustrated and determined, I embarked on a journey that

ventured beyond the conventional. I sought the wisdom of ancient healing practices and more Eastern approaches. I found myself in the offices of acupuncturists and Reiki practitioners, willing to explore alternative paths in pursuit of relief. While their treatments provided moments of respite, the underlying issues continued to linger, leaving me in a perpetual state of searching.

As I explored holistic healing, I discovered a vast tapestry of unconventional therapies and ancient traditions. From Ayurveda to energy healing, I delved into the ideas of diverse modalities, looking for anything that could bring my health back. Each encounter offered a glimmer of hope, but ultimately came up short.

If you've never dealt with this all too common experience of mystery symptoms and multiple perplexed doctors, it is full of fear, despair, and confusion. I would liken it to being in an ocean without a life jacket, flailing about, while the people around you in the water are comfortably wading around with their life jackets, oblivious to your situation. The sad reality is that there is a large part of the population quietly going through this.

One summer day in 2019, walking back to my NYC apartment in a fog of fatigue, I passed a wellness center with infrared saunas. A few weeks earlier, a friend had off-handedly suggested saunas for health. I decided to try it.

I walked in, enjoying the cool air-conditioned interior. I was led into a private room, where a wooden box with a clear, frosted door sat. It looked simple enough, but I couldn't help but feel a sense of apprehension. Was I really about to put myself in a heated box? Could this really help me?

They politely said to get undressed, bring in a couple towels, and sit in there for 30 minutes or so. They also mentioned, "If you feel lightheaded at all, don't hesitate to step out and grab some water before going back in when you feel better." I gave a resigned nod, convinced that I was about to add yet another item to the growing list of failed remedies.

I laid a white cotton towel down on the wooden bench, sat down, and began to feel the heat from all sides. Within 5 minutes of being inside, I began to sweat. Then the sweat picked up. Then the sweat began pouring out of me.

I'd been to exercise classes and sweat heavily before, but this felt like something different. Every time I wiped the sweat with a towel, new sweat would appear within seconds. Wipe, more sweat. Wipe, more sweat. Wipe, more sweat.

30 minutes and 3 sopping wet towels later, I stepped out and took a shower in the clean white restroom adjacent to the sauna room. And that's when the results hit me.

I got out of the shower feeling better than I'd felt since I was a kid. My brain felt like it was more fully online, without the brain fog or sluggishness I had grown so used to. My vision felt improved. Colors were undoubtedly more vivid. I felt a level of energy that simply hadn't been there.

It was a moment of revelation. I stood there, staring at my reflection in the mirror, and I could hardly believe the shift. I felt lighter, clearer, more alive. It was as if a veil had been lifted, and I was seeing the world in high definition for the first time.

I remember thinking to myself, 'Is this how I'm supposed to feel?' It was a profound realization, one that sent a shiver down my spine. Sweat, something so simple, so natural, had given me a glimpse of a new level of well-being.

What the heck happened? How did *sweating* help me more than all the other doctor recommendations COMBINED? Was this experience unique to me? Or did others experience this profound of an impact?

That single session helped reduce my symptoms temporarily. I found that with regular use, I was able to keep lowering and lowering them. Over time, I went from feeling terrible, to normal, to incredible. It felt like I was inching my way to my best self.

This is what sent me down the rabbit hole. I became obsessed. I went home and started researching sweat, health, and saunas. I read books, spoke with experts, attended lectures, and gathered testimonials. I needed to understand why this was happening and to understand if others had the same experience.

When I stepped out of my first sauna session, I couldn't get a question out of my head: What does "true health" feel like?

We each live in the solitude of our own minds, our own bodies. We have no benchmark for comparison, no way to truly know if our baseline of mood, energy, and mental clarity is the same as everyone else's. We only get a glimpse of this comparison when we fall ill or endure a hangover. Suddenly, we yearn for the comfort of 'normalcy'.

But what if our understanding of 'normal' is incomplete? What if there are undiscovered aspects of our health and well-being that could elevate our daily experience? What if our moment-to-moment existence could be enhanced, not through extraordinary means, but through tapping into the power of our bodies?

That sauna session opened my eyes to this possibility. It made me question the boundaries of my own experience, and it ignited a curiosity that I couldn't ignore. It was the beginning of a journey, a quest to redefine 'normal' and to unlock the full potential of true health.

The goal is simple. To maximize our energy levels, mental clarity, mood, and vitality for this short time we have here on this planet. When these aspects of our health aren't performing, our ability to appreciate the magic of our world, the beauty of our experiences, the gift of getting to take action, and so much more isn't there. It's like clouds covering the sun.

Personally, I believe most people don't realize how much better they can feel and how much sweat is an indispensable part of the process. In this book, I'm excited to share with you why.

We'll explore the mythology, history, and science behind sweat and how it is a primary way for our body to eliminate toxins and pollutants. We'll delve into the various methods of sweating, including saunas and exercise, and how they can be used to drastically improve health and well-being. And we'll discuss the utterly transformative benefits people experience when they incorporate more sweating into their lives. I wrote this book because I couldn't find everything I had learned in one place. My hope is that you take the principles in here to heart, apply them, and share them with your community.

Our lives are intrinsically tied to our energy levels, our mental clarity, and our vitality. These elements shape our experiences, our relationships, our achievements, our destinies. Perhaps, like me, you've found yourself backed into a corner by health challenges. Or perhaps you're grappling with a persistent, nagging feeling that something isn't quite right with your health or emotions.

Maybe you're even questioning the societal norms of health and well-being, sensing that there's more to the story than you've been told. Whether you're dealing with a clear health challenge or simply suspect that there's a more vibrant state of 'normal' health to be achieved, this journey is for you.

Wherever you are, whatever the case, this book, and sweat, is for you.

By the end of this book, you'll have a greater understanding of the crucial role that sweat plays in creating optimal health and,

I hope, be inspired to apply what you learn.

What I found was remarkable. I developed a new lens for viewing health and disease, a new appreciation for the intricacy of the human body, and a far deeper understanding of the mental and physical health issues facing modern humanity.

Dive Deeper into Sauna Wellness with Your Exclusive Sauna Mastery Guide

SAUNA GUIDE

A complete handbook for sauna use. Learn the
actionable steps to making sauna use a part of your
wellness routine. Transform your health, vitality, and
longevity

Justin Glaser

In addition to this book, readers can **download a free, 20-page guide** on sauna use, which includes:

· Detailed breakdown of sauna types
· Expert guidance on sauna sessions
· Powerful advice on diet and supplementation
· Pitfalls to avoid with saunas
· Finding a sauna near you
· How to combine sauna use with other wellness practices
· Digestible, actionable steps and practical tips

This free guide is designed to give you everything you need to feel confident in making sauna a regular part of your life. We'll cover much of the content in this guide within the book; however, it can be helpful to have separately.

Just go to <u>sweatthebook.com/saunaguide</u> to get yours now.

2

Sweat History

Sweat. When you hear that word, what images and thoughts does it evoke in your mind? Perhaps you think of exhaustion after a long day, the aftermath of an intense workout, an unpleasant smell. These are common "sweat associations" that many of us have today.

As you'll soon discover, sweat is far more than our culture has led us to believe.

In today's world, we often dismiss sweat as a mere inconvenience or a simple sign of a good workout. Yet, as I navigated the annals of human history, I was struck by a fascinating revelation - across the globe and throughout time, sweat has been revered as a symbol of purification and transformation. It has been woven into the fabric of our mythologies, our religious practices, and our understanding of the human condition.

Our ancestors recognized the sacredness of sweat, its life-giving properties, and its integral role in our existence. This understanding reshaped my perception of sweat and its role in our health and well-being.

I invite you to join me on this exploration of sweat's history, a journey that takes us back in time, tracing the depiction of sweat from ancient mythologies and sacred texts to its role in modern-day rituals. It's a journey that reveals the gap between our historical connection to sweat and how we perceive it today, urging us to reevaluate our relationship with this vital life force.

Norse Mythology: The Creation of Dwarfs

The creation of dwarfs is a fascinating tale that underscores the significance of sweat in the process of creation.

In the farthest reaches of the north, where the aurora borealis paints the sky with ethereal hues and the winters are as relentless as they are long, a tale has endured. It's a tale of sweat, of creation, of the birth of the dwarfs.

Picture the primordial realm of Norse mythology, a world teeming with gods and giants, brimming with magic and mystery. In this realm, a giant named Ymir, a being of immense power, shapes the world around him. As Ymir sleeps one night, the sweat under his left arm coalesces, takes shape, and from it emerges a man and a woman. From the sweat between Ymir's legs, a son is born. These are the first of the dwarfs, beings born of sweat and power, the earth miners, the keepers of the underground.

In the Norse myths, humanity itself was born of sweat, their forms emerging from the perspiration of a primordial giant. Sweat was not just a byproduct of exertion, but the divine essence that breathed life into the cosmos.

A Roman Elixir: The Sweat of Gladiators

In the grand amphitheaters of ancient Rome, sweat took on a value that might seem surprising to us today. Gladiators, the superstars of their time, engaged in fierce combat under the hot Mediterranean sun.

The sweat of gladiators was collected and sold as a cosmetic product. This "gladiator sweat" (translating to gladiatorum sudor) was scraped from the fighters' bodies using a tool called a strigil, a curved metal instrument designed for scraping sweat and dirt from the skin. The sweat, mixed with oil and dust from the arena, was then collected and bottled.

Why was this sweat so prized? It was believed to have powerful properties. Women used it as a beauty treatment, applying it to their skin in the belief that it would enhance their complexion and make them more attractive. It was also thought to have aphrodisiac properties, making it a popular ingredient in love potions.

This practice might seem strange to us today, but it speaks to the cultural significance of sweat in ancient Rome. Sweat was not seen as something to be avoided or concealed, but as a symbol of strength, courage, and sexual allure. It was a physical manifestation of the gladiators' prowess, a tangible reminder of their heroic feats in the arena.

In our modern world, we often overlook the importance of sweat, seeing it as nothing more than an inconvenient byproduct. Yet, these ancient stories hint at a new attitude towards sweat. It's a wake-up call for us to reevaluate our relationship with sweat, to appreciate it not just as a physiological response, but as an integral piece of our human story. As we'll learn later on in this book, it's a critical part of our physical, mental, and spiritual wellbeing, too.

William Bratton: A Lewis and Clark Sauna Story

In the early 19th century, a towering figure named William Bratton, a blacksmith by trade and a member of the Lewis and Clark Expedition, experienced a profound healing journey that underscored the power of sweat and sauna. This tale, woven into the fabric of American history, offers a compelling testament to the therapeutic potential of these ancient practices.

Bratton, a man of few words but immense resilience, was a vital part of the Corps of Discovery, a specially established unit of the United States Army for the Lewis and Clark Expedition. His life took a dramatic turn during the expedition when he fell gravely ill in 1806. His ailment, characterized by debilitating lower back pain and weakness, rendered him incapable of sitting or walking. Despite the best efforts of the expedition's captain-doctors, who administered various treatments including cinchona bark and laxatives, Bratton's condition showed no signs of improvement.

His suffering continued for four agonizing months, during which his pain was mentioned in the captains' journals on 23 different occasions. As the Corps journeyed home up the Columbia River, Bratton's condition remained dire. His weakness was so profound that he had to ride a horse as he was unable to walk.

However, a turning point in Bratton's health saga came on May 24, 1806, at Camp Chopunnish. John Shields, the Corps' principal blacksmith, suggested a radical treatment: inducing "violent sweats" through a primitive form of sauna. This involved digging a four-foot-deep hole, heating stones at the bottom with fire, and then placing Bratton, naked, on a seat above the hot stones. Water was sprinkled on the stones to produce steam, and Bratton was covered with a makeshift tent of willow poles and blankets.

For 20 minutes, Bratton endured the intense heat, drinking copious amounts of horse mint tea. He was then lifted out and plunged twice into the nearby cold river. After another round of steam-induced sweating, he was wrapped in blankets and allowed to cool off gradually.

The results were nothing short of miraculous. The following day, Bratton was able to walk around the camp, nearly free of pain, and began to regain his strength. Two weeks later, he was declared cured. This remarkable recovery, believed to be from a severe case of degenerative disc disease and accompanying arthritis, underscored the healing power of sweat and sauna.

Sweat in Practice: Native American Sweat Lodge

It's the early 1500s in what is now known as the American Southwest. As you wander through the arid landscape, you sight a small Native American village in the distance. As you approach the village, you see a small hut made of adobe bricks with smoke rising from the top. The ground around the hut is charred from years of use.

As you draw closer, you hear soft chanting coming from inside. You push the heavy hide door aside and enter the dimly lit space. The heat hits you like a wave. You see a group of people gathered inside, sweating profusely. They are chanting and singing, their faces contorted in pain and concentration.

This is a traditional sweat lodge, a purification ceremony used by many Native American tribes. The heat of the lodge represents the womb of Mother Earth, and the sweat is believed to cleanse the body and connect the participants to the divine.

As you observe the ceremony, you are entranced by the religious level of commitment to this sweating ritual. Why are they doing this? What purpose are they drawing from the experience?

To the Native Americans, sweating was more than just a way to stay clean. It was a sacred practice that had been passed down through generations. The sweat lodge was seen as a spiritual place, a place where people could connect with the divine and with each other.

The sweat lodge itself was typically made of saplings or willow branches, which were bent and woven together to form a dome-shaped structure. The outside of the lodge was then covered with hides or blankets, and a small hole was left at the top to allow smoke to escape. Inside the lodge, heated rocks were placed in a pit, and water was poured over them to create steam. The participants, often led by a shaman or medicine man, would then enter the lodge and sit in a circle around the pit.

The experience of sweating in the lodge was intense. The heat and humidity were overwhelming, and the participants would often feel as though they were suffocating. But this was all part of the experience. The intense heat was seen as a way to purify the body, and the darkness of the lodge was seen as a way to focus the mind and connect with the spirit world.

The sweat lodge ceremony was often accompanied by singing, chanting, and the burning of sage or other herbs. The participants would often pray for healing, guidance, or protection, and would offer thanks to the spirits for their blessings.

The intense heat, the rhythmic chants, the communal energy— all contribute to a phenomenon that seems to tap into deeper layers of human consciousness, a mystery that beckons modern science to unravel its secrets.

Archaeological evidence suggests that sweat lodges were used by Native American cultures thousands of years ago. Remains of ancient sweat lodges have been found across North America, from the southern parts of Canada to the southwestern United States.

One of the oldest known sweat lodges was discovered in Saskatchewan, Canada, dating back to around 5000 years ago. Similarly, in the southwestern United States, the Ancestral

Puebloans, a Native American culture that flourished from approximately AD 100 to 1600, are known to have used sweat lodges.

Scientific Mysteries: Unveiling the Healing Secrets

Immune System Boost

In various Native American narratives, it is often recounted how individuals who frequently participated in sweat lodge ceremonies exhibited robust immune systems. This phenomenon, akin to the benefits of modern-day saunas, is a testament to the ancient wisdom embedded in these practices.

Healing Skin Ailments

The sweat lodge ceremonies were known to have a profound effect on skin health. Individuals with skin ailments often found relief after participating in these rituals. The combination of heat and medicinal herbs used in the heat helped in alleviating skin conditions like psoriasis and eczema, offering a natural remedy that harnessed the healing powers of the earth.

Alleviating Rheumatic Diseases

Native American elders often recount how the sweat lodge ceremonies provided relief from rheumatic ailments. The heat and humidity inside the lodge helped in easing the stiffness and pain associated with conditions like arthritis. It was a natural therapy that facilitated mobility and reduced discomfort, enhancing the quality of life for many.

Emotional and Psychological Healing

The sweat lodge was a sanctuary for those grappling with emotional and psychological challenges. Many Native American tribes believed in the therapeutic potential of these ceremonies to heal the mind. Individuals battling depression, anxiety, and trauma found solace and healing within the sacred space of the lodge, a place where they could release their burdens and find emotional equilibrium.

Native Americans intuitively understood the impact of sweat and sauna on their mental, physical, and spiritual health. In future chapters, we'll dive deeper into how science has caught up to their innate understanding.

Native American Sweat Lodges: A Test of Endurance and Spiritual Discipline

Sweat lodges have been a significant part of Native American cultures for centuries, serving as sacred spaces for purification, prayer, and healing. These ceremonies are deeply spiritual and often seen as a test of one's physical endurance and spiritual discipline.

In many tribes, sweat lodge ceremonies are considered rites of passage, marking significant transitions in life. Young boys might participate in a sweat lodge ceremony as part of their transition into manhood. The ability to withstand the intense heat and steam inside the lodge is seen as a testament to their strength and resilience, preparing them for the challenges they will face as adults.

In fact, the protagonists in many traditional Native North American Tales are tested in sweat lodges.

But the sweat lodge is not just a test of physical endurance. It's also a test of spiritual discipline. Participants are encouraged to enter the lodge with a clear intention or prayer. The intense heat and steam serve to purify not just the body, but the mind and spirit as well. As participants sweat, they are encouraged to let go of any negative energy or thoughts, focusing instead on their prayers and intentions.

The ceremony is led by a trained facilitator, often a tribal elder or spiritual leader, who guides the participants through the process. The facilitator ensures the safety of the participants, controls the heat inside the lodge, and leads the group in prayer and song. The ceremony often involves multiple rounds of heating and cooling, each round representing a different direction, element, or stage of life.

Despite the physical discomfort, participants often report feeling a sense of peace and clarity after a sweat lodge ceremony. The process of sweating, praying, and enduring the heat together can create a sense of community and shared experience. It's a time for reflection, connection, and spiritual growth.

The sweat lodge ceremony is a powerful tradition in Native American cultures. It's a practice that goes beyond mere physical cleansing, offering a path to spiritual purification, self-reflection, and communal bonding. Through the sweat lodge, participants learn to face challenges with strength and resilience, carrying these lessons with them as they navigate the journey of life.

Sweat lodges like the Native Americans used have often been linked to creation itself. Leaving the sweat lodge is often associated with being "reborn".

Scandinavian Saunas: A Tradition of Wellness and Community

The Scandinavian cultures, particularly the Finns, have a long-standing tradition of using sweat lodges, known as saunas, for both physical and social well-being. The sauna in Finland is an old phenomenon, and its origins are shrouded in the mists of time. Some archaeological evidence suggests that the use of saunas in Finland dates back over 2000 years.

The traditional Finnish sauna is a small wooden room or house designed to hold heat, typically from a wood-burning stove topped with stones. Water is thrown onto the heated stones to produce steam, which increases the humidity and heat within the sauna, causing the occupants to sweat profusely.

Saunas in Finland are not merely a luxury; they are a way of life. It is said that in Finland, there are more saunas than cars. They are found in homes, offices, and even at the Finnish Parliament House. The sauna is a place for relaxation, contemplation, and above all, for cleansing the body and mind.

In the warmth of the sauna, Finns sweat while discussing everything from personal problems to politics. It is a place where social hierarchies are forgotten, and everyone is equal.

The sauna is often followed by a cooling period, which can involve a swim in a nearby lake or sea, regardless of the season.

In addition to their use in everyday life, saunas also play a role in special occasions in Finland. They are often used in conjunction with holidays, family gatherings, and even business meetings. Some Finns also observe the tradition of "löyly," a ritual that involves gently whipping oneself with a bundle of birch twigs while in the sauna, believed to stimulate the skin and promote circulation. The word löyly comes from an old Finnish word that means spirit or life. This reflects the belief in the healing and rejuvenating properties of the sweating sauna experience

The sauna is a deeply ingrained part of Finnish culture, serving as a place for physical cleansing, relaxation, and social bonding. The tradition of the sauna in Finland is a testament to the Scandinavian appreciation for wellness, community, and a connection with nature.

Sweating Around The World

The use of sweating for health and spiritual purposes is not unique to the cultures we've explored so far. In fact, it has been used by dozens of different cultures around the world for thousands of years. From the ancient Greeks and Romans to the Maasai people of East Africa, the practice of sweating has been revered as a way to cleanse the body and connect with the divine.

The ancient Greeks and Romans used sweating as a form of

relaxation and rejuvenation. They built elaborate bathhouses and steam rooms.

In Japan, the traditional practice of "Mushiburo" involves sitting in a small hut filled with steam from hot rocks, which they believe promotes overall health and well-being.

In Africa, the Maasai people use a sweat lodge called an "Enkang," which is believed to have healing properties and which they have used to treat a variety of ailments.

In India, Ayurvedic medicine has long used sweating as a way to eliminate toxins from the body and improve overall health. The practice of "svedana" involves sitting in a steam box filled with aromatic herbs.

And in many indigenous cultures in Central and South America, the sweat lodge is a central part of spiritual and healing practices, used to cleanse the body and connect with the divine.

Today, science has caught up to what all these disparate cultures intuitively knew: that sweating is a profound, powerful way to promote healing, health, and well-being.

In this journey through time and across cultures, we've seen how sweat has been revered, utilized, and celebrated. From the creation myths of Norse mythology to the purification rituals of Native American sweat lodges, from the beauty treatments of ancient Rome to the communal bonding in Finnish saunas, sweat has been an integral part of human life. It's been a symbol of strength, a tool for purification, a catalyst for transformation, and a pathway to the divine.

Yet, in our modern world, we've distanced ourselves from this vital life force. We've come to see sweat as an inconvenience, something to be hidden away. But as we delve deeper into the science of sweat and its role in our health and well-being, we begin to see the wisdom of our ancestors in a new light.

In the next chapter, we'll delve into the science of sweat, exploring how it works and why it's so crucial for our health and well-being. As we do, remember the stories we've shared here, and consider how our understanding of sweat has evolved over time.

Key Takeaways

The Multifaceted Role of Sweat: Throughout history, different cultures have revered sweat for its various roles, including its significance in creation myths, as a mark of heroism, and as a medium for spiritual connection and healing.

   Norse Mythology and the Creation of Dwarfs: In Norse mythology, the creation of dwarfs is intricately linked to sweat, symbolizing the divine essence that breathed life into the cosmos. This narrative showcases sweat as a vital force in the process of creation, hinting at its deeper significance in human life.

   The Roman Elixir - Gladiator Sweat: In ancient Rome, the sweat of gladiators was highly prized for its supposed beauty-enhancing and aphrodisiac properties, reflecting the cultural significance of sweat as a symbol of strength, courage, and sexual allure.

Native American Sweat Lodges: The Native American sweat lodge ceremonies represent a profound spiritual practice, facilitating a deep connection with the divine and fostering community bonding. These ceremonies, rooted in ancient traditions, offer both physical and emotional healing benefits.

Scientific Mysteries and Modern Relevance: The chapter hints at the scientific mysteries surrounding the benefits of sweat, encouraging readers to delve deeper into understanding the physiological and psychological advantages that sweating can offer in modern times.

A Call to Rediscover the Significance of Sweat: The chapter serves as a wake-up call for readers to rediscover the multifaceted significance of sweat, urging them to embrace it as a critical aspect of physical, mental, and spiritual health.

3

How Sweat Works

In our modern existence, we often overlook sweat's profound importance, relegating it to a mere side effect. But now, we're going to explore the intricate workings of sweat—a journey that will reveal the remarkable depths of this seemingly mundane bodily function.

Dr. Jonas Salk, a distinguished physician and visionary medical researcher known for his pivotal role in developing the polio vaccine, stood in awe of the human body. He called it "a complex and wondrous creation, its mechanisms of homeostasis and self-regulation a testament to the brilliance of nature."

He recognized the body's complexity, the delicate dance of its interconnected systems, and the genius of its self-regulating abilities.

By gaining a deeper understanding of how sweat works, we gain a newfound appreciation for the miraculous mechanisms that contribute to our health, vitality, and abilities to prevent disease.

Fever Dream

"Give me a fever, and I can cure any disease." – Hippocrates

Imagine this: you're tucked into bed, your body wrapped in a cocoon of blankets. Outside, the night is cold and still, but inside, you're burning up. Your forehead is hot to the touch, your body aches, and you're shivering despite the heat. You're in the throes of a fever, your body's response to an invader, be it a virus or bacteria.

In this state, your body is a battleground, and your immune system is the frontline soldier. It raises your body's temperature to create an inhospitable environment for the invaders. But this increase in temperature comes with a cost. Your body, a finely tuned machine, operates best at a specific temperature, around 98.6 degrees Fahrenheit (37 degrees Celsius). When your temperature rises, your body needs to cool down, and that's where sweat enters the picture.

Sweat, that liquid that beads on your forehead during a fever or a strenuous workout, is more than just water. It's a cocktail of water, salt, and other electrolytes, along with urea, ammonia, and other waste products. It's your body's built-in air conditioner, a marvel of human biology that's been keeping us cool for millions of years.

As your fever rages on, your hypothalamus, the part of your

brain that acts as your body's thermostat, sends a signal to your sweat glands. You have between 2 to 5 million of them, scattered across your body, ready to spring into action at a moment's notice. They start producing sweat, which then seeps out onto the surface of your skin.

The moment the sweat hits your skin, it starts to evaporate. This evaporation process, as simple as it may seem, is actually a complex physical phenomenon that cools your body down. As the sweat evaporates, it takes some of your body's heat with it, dissipating it into the air. This helps to lower your body's temperature, bringing it closer to its normal state.

In ancient Greek and Roman times, sweating during a fever was often seen as a purifying process. The Greek physician Hippocrates, known as the father of medicine, believed that fevers and subsequent sweating were the body's way of ridding itself of harmful substances. This concept aligns with the idea of "cleansing" the body through sweating, suggesting that fever-induced sweating was considered a positive sign of the body's healing process.

In traditional Chinese medicine, sweating during a fever was also viewed as a means of expelling toxins from the body. Sweating was believed to open the body's pores and release pathogenic influences, allowing for the restoration of balance and health.

Before we dive into what sweating does for us, let's take a step back. How does sweating work?

It's something you (hopefully) do quite often, but have probably given very little thought.

Remember those millions of sweat glands? Sweat glands are classified into two types – eccrine and apocrine. Eccrine sweat glands are present across the body and are responsible for producing sweat to regulate body temperature. These glands are most densely packed on the palms of our hands, the soles of our feet, and our forehead, with as many as 370 to 700 sweat glands per square centimeter. Apocrine sweat glands, on the other hand, are found in specific areas of the body, such as the armpits, the groin, and around the nipples. Unlike eccrine glands, which are active from birth, apocrine glands only become active during puberty.

Apocrine sweat glands are larger than eccrine glands and are connected to hair follicles. They produce a thicker, milky sweat that contains proteins and lipids. This sweat is initially odorless, but when it comes into contact with the bacteria on the skin's surface, it can produce a distinctive body odor. This is why areas with a higher concentration of apocrine glands, like the armpits, are often associated with body odor.

Interestingly, apocrine sweat glands are also believed to produce pheromones, chemicals that can subtly influence the behavior or physiology of others. This is thought to be a remnant of our evolutionary past, where such signals played a crucial role in communication between individuals.

In addition to their role in thermoregulation and communication, apocrine sweat glands also play a part in skin health. The sweat they produce helps to moisturize the skin and maintain its pH balance, contributing to the skin's barrier function.

While women tend to have more eccrine sweat glands overall, men's glands are often more active and produce more sweat. This is largely due to differences in body size, muscle mass, and hormones. Men typically have a larger body mass and more muscle mass than women, which leads to higher metabolic rates and, consequently, more heat production. This heat needs to be dissipated, and one of the main ways the body does this is through sweating.

On average, a person can produce between 0.8 to 1.4 liters of sweat per hour during intense exercise, but this can vary greatly depending on factors like heat, humidity, and individual fitness levels.

What's in sweat? Sweat is composed of water, salt, and other electrolytes such as sodium, potassium, and magnesium. As you know by now, it also contains urea, ammonia, and other waste products that are excreted through sweat. In addition, sweat can contain heavy metals, such as lead and mercury, and other environmental toxins, such as BPA and phthalates, that can accumulate in the body over time. This is an absolutely crucial point which, I believe, has been massively overlooked in the health community. We'll be exploring this in far greater depth starting in Chapter 5.

The process of sweating begins when the body's internal temperature rises above its set point. This can happen due to physical activity, exposure to hot temperatures, or an increase in the body's metabolic rate. In response, the hypothalamus, a part of the brain that regulates body temperature, signals the sweat glands to begin producing sweat.

As sweat is produced, it is secreted onto the surface of the skin, where it evaporates and cools the body down. The evaporation of sweat is a key part of the cooling process, as it helps to dissipate heat from the body. Sweat also helps to reduce body temperature by increasing blood flow to the skin, where heat can be dissipated.

Sweat Functions Across The Animal Kingdom

Sweating is a remarkable phenomenon found in various species, each with its own unique adaptations to cope with heat stress and regulate body temperature. But, as we'll see, nobody does it quite like we humans do. While humans are known for their profuse sweating, let's take a look at temperature regulation techniques across different animals and discover how they cool down in their own extraordinary ways.

Imagine a hot savannah where a majestic lioness prowls under the scorching sun. Unlike humans, lions don't have sweat glands covering their entire body. Instead, they rely on other cooling mechanisms. As the lioness rests in the shade, her powerful chest rises and falls with each breath. When heat builds up, she extends her tongue and pants heavily, allowing the moist surface to evaporate, dissipating body heat. This panting behavior helps cool her down, ensuring she's ready for action when the time comes.

In a desert landscape, you might find a spiky-tailed iguana basking on a rocky outcrop. Unlike mammals, reptiles lack sweat glands altogether. To regulate their body temperature, they rely on external heat sources. Our iguana friend soaks up the sun's rays, absorbing warmth into its scaly body. When it becomes too hot, it seeks shelter in a burrow, where the cool underground environment provides relief from the sweltering heat.

In the avian realm, we encounter the charismatic hummingbird, known for its vibrant colors and rapid wingbeats. These tiny birds don't sweat like humans do, but they have their own tricks for staying cool. On a blazing summer day, a ruby-throated hummingbird hovers near a flower, sipping nectar with its long, slender beak. As it feeds, the hummingbird's metabolic rate increases, generating heat. To prevent overheating, it rapidly flaps its wings, creating a miniature breeze that helps dissipate excess warmth.

As we venture into the insect world, we encounter the industrious honeybee. These tiny creatures work tirelessly, buzzing from flower to flower in search of nectar. Despite their small size, honeybees face temperature challenges too. To combat heat stress, bees employ a fascinating technique called "bearding." On hot days, thousands of bees gather near the entrance of the hive, forming a cluster that resembles a fuzzy beard. By huddling together, they create airflow within the hive, allowing heat to escape and maintaining a cool, comfortable environment for their colony.

Comparing these stories to our own sweating experiences, we realize the diverse ways animals have adapted to survive in their respective environments. While humans have a higher density of sweat glands, allowing us to engage in intense activities and withstand higher temperatures, other animals have developed alternative thermoregulatory strategies suited to their unique physiological and ecological needs.

One thing is clear amongst evolutionary biologists: humans' ability to sweat played a key role in our ability to survive and thrive on planet earth. Sweating is a highly efficient cooling mechanism that has allowed humans to thrive in a variety of climates and to engage in endurance activities, such as long-distance running, that other species cannot. This ability to regulate body temperature through sweating has allowed our ancestors to hunt and gather food over long distances, to migrate across diverse terrains, and to adapt to changing environments. Furthermore, the development of eccrine sweat glands, which are unique to humans and produce the watery sweat that cools us down, coincided with the evolution of other distinctly human traits, such as reduced body hair and increased brain size. Evidently, sweating is not just a physiological response, but a fundamental part of our evolutionary history.

Nerves Getting To You? Emotional Sweating

In the realm of sweating, we have delved into the fascinating world of thermoregulatory perspiration, where droplets form on our brows and trickle down our backs to cool our fevered bodies. But what about another kind of sweating, the kind that pours forth in moments of emotional intensity or when the weight of stress bears down upon us?

Emotional sweating is a curious phenomenon. It is a silent response, hidden from the casual observer, yet palpable to those who experience it. Picture this: you find yourself in a nerve-racking situation, perhaps standing before a crowd about to deliver a speech, or sitting across from someone you desire deeply. Suddenly, your palms become moist, your underarms dampen, and a subtle sheen graces your forehead. It is the manifestation of emotional sweating, a physical reflection of the intensity coursing through your veins.

But how does this process unfold within our bodies? Emotional sweating is orchestrated by the sympathetic nervous system, the part of our autonomic nervous system responsible for our fight-or-flight response. When we encounter emotional triggers, be it anxiety, fear, or exhilaration, the sympathetic nervous system leaps into action, like a conductor guiding a symphony of sweat.

Deep within our skin, nestled among a vast network of nerves, lie the eccrine sweat glands.

These remarkable structures, found throughout the body, possess a unique ability to produce sweat in response to neural signals. When the sympathetic nervous system receives the cue, it sends messages to these sweat glands, commanding them to release their watery cargo. And so, a cascade of emotional sweat ensues, leaving its mark on our skin.

Why does our body respond in this way? What purpose does emotional sweating serve?

Firstly, emotional sweating is thought to be a means of communicating non-verbally with others. In a social context, the smell of sweat can convey important information about an individual's emotional state. It can signal danger, fear, or even arousal, allowing others to react and respond accordingly.

Secondly, emotional sweating is believed to have played a role in thermoregulation during moments of heightened physiological activity. When our bodies prepare for a fight-or-flight response, there is an increase in metabolic activity, leading to an elevation in body temperature. Sweating helps dissipate heat, cooling down the body and preventing overheating during these intense emotional states.

Additionally, some researchers suggest that emotional sweating may have played a role in enhancing grip and traction during physically demanding situations. The moisture from sweat can improve the friction between our skin and objects we interact with, providing a better grip for survival-related tasks.

While the precise mechanisms and functions of emotional sweating are still being explored, it is clear that our bodies have retained this ancient response as a part of our physiological repertoire.

Practical Advice

So what are your choices for sweat? In our exploration of the many ways our bodies can sweat, one method stands out as the most powerful tool for unlocking a multitude of health benefits: the sauna. From the Native American sweat lodges to the Russian banyas, our ancestors were onto something. Today, science backs them up.

While various activities like hot yoga, exercise, and even hot baths can induce sweat, the sauna offers a unique and concentrated experience that maximizes the potential benefits. We'll dive deeper into this in the coming chapters.

Nonetheless, below are some brief overviews of different ways you can incorporate regular sweat into your life.

1. **Exercise**: One of the most common ways to sweat is through exercise. When we engage in physical activity, our bodies produce heat, which causes us to sweat. This can help to eliminate toxins and waste products from the body, and it also has a number of other health benefits, including improved cardiovascular health, increased muscle strength, and improved mental health. Whether through sports, weight lifting, sex, running, or exercise classes, exercise is an excellent way to incorporate sweating into your life.

2. **Sauna Therapy**: Sauna therapy involves sitting in a heated room, typically between 140°F and 200°F, for a period of time. This can cause us to sweat heavily, which helps to eliminate toxins and other harmful substances from

the body. In addition to its detoxifying effects, sauna therapy has also been shown to have a number of other health benefits, including improved cardiovascular health, reduced inflammation, and improved immune function.

3. **Hot Yoga**: Hot yoga is a type of yoga that is practiced in a heated room, typically between 90°F and 105°F. This can cause us to sweat heavily, which helps to eliminate toxins from the body. Hot yoga can also help to improve flexibility, strength, and balance.

4. **Steam Rooms**: Steam rooms are similar to saunas, but they use moist heat rather than dry heat. This can cause us to sweat heavily, which helps to eliminate toxins from the body. In addition to its detoxifying effects, steam rooms can also help to improve respiratory health and reduce stress and anxiety. Compared to a dry sauna or infrared sauna, steam rooms are the least effective option and should probably be avoided. Typically, the water quality is questionable (tap water almost always contains toxins) and you're then breathing in the moisture, straight into your lungs. Not great.

5. **Hot Baths**: Taking a hot bath can also help to stimulate sweating and eliminate toxins from the body. This is because the hot water causes the body to produce heat, which can cause us to sweat. Taking a hot bath can also help to reduce stress and improve relaxation.

Key Takeaways

· **The Complexity of Sweat**: Sweat, a mix of water, salt,

electrolytes, and waste products, is our body's natural cooling system.

- **Sweat and Fever**: Sweat plays a crucial role in cooling down the body during a fever.

- **Sweat Glands**: Humans have 2 to 5 million sweat glands of two types - eccrine and apocrine - spread across the body.

- **Sweat Across the Animal Kingdom**: Different species have unique adaptations for heat stress and body temperature regulation.

- **Emotional Sweating**: Emotional triggers like anxiety, fear, or exhilaration can cause a sweat response.

- **Sweating as a Detoxifying Agent**: Regular sweating can help eliminate heavy metals and other toxins from the body.

4

Toxins: A New Perspective

Before we get into the science of sweat and detoxification, it's critical for us to understand the omnipresence of toxins in our everyday lives.

Imagine the miracle of a newborn baby, a symbol of hope and new beginnings. Yet, even at this early stage, a baby has been impacted by the world in which we live.

According to the Environmental Working Group (EWG), a newborn's umbilical cord blood contains an average of 200 chemicals.

Ten infants were randomly selected for this EWG study from the pool of live births during the summer of 2004, all part of the Red Cross's national cord blood collection program. These infants were not specifically chosen due to any known prenatal exposure to harmful chemicals. Yet, astonishingly, each of these newborns entered the world already carrying a wide range of contaminants in their bodies.

Among the total 287 chemicals found, 180 are known to cause cancer in humans or animals, 217 are toxic to the brain and nervous system, and 208 cause birth defects or abnormal development in animal tests. This statistic is deeply troubling, but also serves as a powerful call to action.

In this chapter, we're going to explore a topic often overlooked: toxins. These silent culprits can have profound influences on our health, contributing to the rise in chronic diseases and mental health disorders. But this journey is not about fear. It's about empowerment. It's about arming ourselves with knowledge and understanding so that we can make informed decisions about our health and the health of our loved ones.

By exploring the different categories of toxins, their sources, and their impact on our bodies and minds, we will gain the tools we need to navigate our toxin-filled world. This journey will challenge us to rethink our beliefs about health and disease, and inspire us to take action.

I hope you'll keep an open mind, because it might just change your quality of life. It certainly changed mine.

A New Lens on Health and Wellbeing

What causes disease? What causes brain fog, anxiety, depression? What causes chronic fatigue? What causes autoimmune issues?

Like many, I initially accepted the prevailing views about the causes of disease and conditions like brain fog, attributing them largely to genetics, diet, or unfortunate luck. Like many people, I had accepted these notions without truly questioning their validity. After all, the world is an impossibly complicated place, and it's only natural for us to seek certainty amidst it all. We have to accept many beliefs we've heard from others, rather than reason everything from the ground up.These were the explanations I had often heard, and they seemed to make sense. And, it turns out, these things do have an impact. But they are not the full picture. My personal health challenges forced me down a path of deeper exploration.

It wasn't until more recently, with a humble curiosity and an open mind, that I began to realize just how limited my understanding was.

My exploration led me down a path that I couldn't have foreseen. In the midst of researching sweat and saunas, I came across ideas and theories that initially seemed peculiar, even fringe. Whether in books, or research papers, or youtube dissertations, I would find varying types of discussion around "detoxification". My knee-jerk reaction to the term "toxin" was one of dismissal, brushing it off as the territory of so-called health fanatics. I mean, we weren't living next to nuclear reactors or chemical plants, so what could we possibly have to worry about? It was all too easy to label these ideas as outside the realm of reason.

But something within me urged me to stay open, to listen with a receptive ear, and to engage in conversations with experts in the field of environmental toxins. It was in these conversations that the seeds of doubt began to sprout, and I realized that there was more to the story than met the eye. I was compelled to peel back the layers, to dig deeper into the realm of toxins and their potential impact on our health and well-being.

As I delved into the depths of research, a fascinating world unveiled itself before me. I discovered a body of evidence that challenged the prevailing narratives about health and disease. I encountered dozens of studies that linked toxin exposure to a myriad of health conditions, ranging from chronic diseases to mental health disorders. It became increasingly clear that toxins were not merely a figment of the imagination but a formidable force lurking in our daily lives. And that they were affecting every aspect of our mental, physical, and spiritual wellbeing.

I moved from skepticism to curiosity, and from curiosity to an unnerving belief:

Toxins are not just minor irritants or inconsequential substances; they are silent culprits wreaking havoc on our health and contributing to the staggering rise in chronic diseases and mental health disorders.

The more I learned, the more I realized that the topic of toxins held the key to understanding a wide array of health challenges plaguing our world.

It became apparent that the effects of toxins extended far beyond the boundaries of nuclear reactors and chemical spills. They seeped into our daily lives, infiltrating our homes, workplaces, and our bodies.

So, here's the question:

What if toxins are a contributing, if not primary, cause of most of our mental, physical, and spiritual problems?

In the chapter that follows, we will embark on an exploration of toxins and their profound influence on our health and well-being. Together, we will challenge the preconceived notions and delve into the realm of critical thinking, unearthing the hidden truths that have eluded us for far too long. This is not a journey for the faint of heart, but one that demands our attention, our willingness to question, and our commitment to change. By understanding the problem, we can become empowered to use solutions that dramatically improve our lives. So let's get into it.

What are Toxins?

"Toxins" may sound like a cliche buzzword, but what exactly are they? And why are they such a big deal for our health?

In simple terms, toxins are substances that are reliably harmful to our bodies. They can be found in the air we breathe, the food we eat, and the products we use every day. Unlike medicines (true medicines), which are designed to heal and improve our health, toxins do the opposite - they reliably and predictably cause problems when they enter our bodies.

Our bodies are designed to get rid of toxins as quickly as possible. We have complex systems in place, such as the liver, kidneys, and skin (through sweat!) that work to filter out toxins and eliminate them from our bodies. But when we are exposed to too many toxins, or when our bodies can't keep up with the amount of toxins we are exposed to, problems arise. And, unfortunately, we are *all* exposed to too many toxins.

They have a range of health effects, depending on the type of toxin and the amount of exposure. They can damage our organs, disrupt our hormones, and interfere with our immune system. They can cause everything from headaches, fatigue, and mild anxiety to cancer and neurological disorders.

In short, toxins are everywhere.

To understand the true extent of our exposure to toxins, we need to take a closer look at the six categories of toxins that we are exposed to every single day of our modern lives: industrial toxins, agricultural toxins, household toxins, personal care product toxins, food toxins, and consumer product toxins.

Industrial toxins include heavy metals, pollution, and radiation released by industrial activities. This can include everything from lead and mercury to air pollution and nuclear fallout. Industrial toxins can have a range of health effects, including damage to the nervous system, reproductive system, and immune system.

Agricultural toxins include pesticides, hormones, and herbicides. These toxins are used to grow our food and can end up in our bodies through the food we eat. Pesticides have been linked to a range of health problems, including cancer, birth defects, and neurological damage.

Household toxins include building materials, rugs, paint, and cleaning supplies. These toxins can be found in many common household items and can have a range of health effects, including respiratory problems, allergies, and skin irritation. Many of the building materials used in our homes, such as paint, carpet, and furniture, can release harmful chemicals into the air. And when we use cleaning products and air fresheners, we are adding even more chemicals to the mix.

Personal care product toxins include perfumes, cosmetics, and health and beauty aids. These products often contain a range of chemicals, including phthalates, parabens, and formaldehyde, which have been linked to a range of health problems, including cancer, reproductive problems, and developmental issues.

Food toxins include GMOs, food coloring, artificial flavors, and artificial sweeteners. These additives are often used to make our food more appealing or to extend its shelf life, but they can have a range of health effects, including allergies, asthma, and behavioral problems. The use of pesticides, herbicides, and other chemicals is rampant.

Consumer product toxins include flame retardants in clothing, toys, and blankets, as well as sealants in cookware and other household items. These toxins can be found in many common household items and can have a range of health effects, including hormone disruption, cancer, and developmental problems.

When we turn on the tap, we expect clean water to come out. But the reality is that our tap water is often contaminated with a range of chemicals, including chlorine, fluoride, and heavy metals. And when we shower, we are exposed to these chemicals through our skin, which is an incredibly effective absorber of substances.

Toxins exist and they're everywhere. But how bad are they really?

According to Dr. Joe Pizzorno, a leading expert in environmental medicine: "many diseases once rare are now epidemic in people of all ages for reasons that doctors can't explain - unless we consider toxins." This is a sobering thought, but it's one that we can't ignore if we want to take control of our health.

So where's the data? How much exposure does the average person have, how much of a 'toxic load' do we have in our bodies, and where is the research on its impact on our health?

Toxins and their mind-boggling health implication

When I started researching toxins (which we should really call xenobiotics, but will call toxins for the sake of simplicity), I had no idea of the scope or scale of the problem.

Most of these toxins simply did not exist prior to the mid 20th century. They are a product of human revolutions. But do they actually negatively impact our health?

Oh yes. In fact, they're largely linked with practically every chronic health problem facing the modern world.

Depression, anxiety, obesity, diabetes, cancer, Alzheimer's, autism, chronic fatigue syndrome, bipolar disorder, schizophrenia, and I could go on.

.

In his foreword to The Autoimmune Epidemic, Dr. Douglas Kerr, M.D., Ph.D. professor at Johns Hopkins School of Medicine, proclaimed:

"There is no doubt that autoimmune diseases are on the rise and our increasing environmental exposure to toxins and chemicals is fueling the risk. The research is sound. The conclusions, unassailable."

Let's explore how toxins can influence different health categories.

Mental and Neurological Health

Toxins can contribute to a range of mental health disorders, including depression, anxiety, and mood disorders.

For children, there are clear links of lead, mercury, and arsenic exposure to developmental disabilities. It affects cognition, academic performance, behavior, and attention in negative ways. Multiple studies have shown that reducing lead's toxic load in the body would improve outcomes for individual children.

We opened this chapter with a disturbing statistic on chemicals passed down from mother to child. A 2016 study was conducted to determine if persistent organic pollutants (POPs) could disrupt child neurodevelopment. POPs are highly resistant to breaking down, and they pass through the placenta. The researchers tested 687 mother-children pairs. The results were astonishing and clear. They tested the children at 4 years of age after having measured their levels before birth. Children with "high" HCB concentrations (hexachlorobenzene) demonstrated decreased scores in perceptual performance, general cognitive, executive function, and working memory tests.

HCB was widely used as a pesticide until the 1960s. Although its production and use have been significantly restricted or banned in many countries under the Stockholm Convention on persistent organic pollutants, HCB can still be found in the environment due to its persistence and past widespread use. It was first used as a fungicide in 1945 and was in use for nearly 2 decades before finally being banned.

Lipophilic (fat loving) chemicals enter the body, either through the air we breathe, the food we eat, or the drinks we drink. They then are attracted to fat, and so instead of going through the bloodstream to be digested and excreted, they lodge into our adipose fat tissue. Depending on the toxin, they can stay there for days, years, or decades. Every time we burn energy, small amounts get released back into the body, continuously "poison dripping" us with symptoms.

It's therefore no wonder that "From 1990 to 2010, mental

and behavioral disorders increased by more than 37%, Parkinson's disease increased by 75%, Alzheimer's disease doubled, autism increased by 30% and attention deficit hyperactivity disorder (ADHD) increased by 16% (Murray et al., 2012; Vos, et al., 2012). The increases in many epidemic and pandemic diseases, including neurological disorders, have been attributed to environmental exposures to exogenous toxic chemicals."

"The prevalence of neurological diseases discussed here as well as the other diseases cited have all increased dramatically in the past half century. For example, the prevalence of bipolar disorder and of schizophrenia has increased in the range of 40% from 1990 to 2010 (Murray et al., 2012), the prevalence of Alzheimer's disease is expected to double every 20 years (Mayeux & Stern, 2012). Such an increase can only be explained by environmental consideration and it corresponds to the worldwide increased use of POPs, plastic additives and other chemicals, fossil fuel and the environmental pollution associated with their use and discharge."

To be clear, these studies are claiming that the dramatic rises across Alzheimer's, Parkinson's, autism, ADHD, bipolar disorder, and schizophrenia are all directly being caused by environmental exposures to toxic chemicals.

The brain is made of approximately 60% fat. An avalanche of fat loving toxins have been introduced to the world over the last 50 years. Is it any wonder brain related symptoms and diseases have skyrocketed when you look at it through this lens? Working to get these substances out of my body (and brain) results in remarkable changes:

- Deeper sleep
- Brain fog lifted
- More stable moods
- Faster response times
- Sense of increased intelligence
- Dramatically increased mental clarity

It wasn't that I became superhuman. It's that I felt like I got back what the toxins had robbed me of. My brain literally had less toxins stored in the fat tissue. In short, less toxins in the brain = better brain function. It's hard to explain the significance of how this before and after felt for me.

I believe this applies to everyone. The default state on the planet today is having a high toxic load that is affecting cognitive performance. Let that sink in.

We'll dive deeper into how this can work for you in the following chapters, so hang tight.

Cancer

We've discussed what was found in newborn's umbilical cord blood already. When it comes to cancer, the evidence is clear. In California, law through Proposition 65 required the state to publish a list of chemicals known to cause cancer, birth defects, or reproductive harm. First published in 1987, it is required to be updated at least once a year. It now includes over 900 chemicals known to cause cancer.

Over 1,400 substances and collections of chemicals are identified or presumed to be carcinogenic. Due to their presence in industrial processes, consumer items, and our daily consumption of food, water, and air, individuals in the United States and around the world encounter these compounds that promote cancer on a daily basis. These hazardous elements infiltrate and accumulate in our bodies.

Reproductive Health

Toxins can interfere with our reproductive health, leading to fertility issues and developmental problems in children. Endocrine-disrupting chemicals, for example, can interfere with hormone function and affect fertility. Let's look at one of the most dramatic examples facing the world today: Atrazine.

In the world of agrochemicals, few substances have sparked as much controversy as Atrazine. Developed by the Swiss company Syngenta, Atrazine quickly became one of the most widely used herbicides in the world, particularly in cornfields across the United States. Its ability to control a broad spectrum of weeds made it a favorite among farmers.

However, questions began to arise about Atrazine's potential impact on health and the environment. In response to these concerns, Syngenta hired Tyrone Hayes, a professor of Integrative Biology at the University of California, Berkeley, to study the herbicide's effects on amphibians. This soon backfired on them.

Hayes, a world-renowned herpetologist (studying amphibians and reptiles), was known for his work on amphibian endocrinology. He embarked on the research, expecting to find little to no impact of Atrazine on amphibians. However, what he discovered was startling.

In his experiments, Hayes exposed African clawed frogs to Atrazine. He found that even at incredibly low levels, atrazine had a dramatic effect on the frogs' reproductive systems. Male frogs exposed to the herbicide developed female reproductive organs, and some even produced eggs. Hayes hypothesized that atrazine was acting as an endocrine disruptor, interfering with the frogs' hormonal systems. Hayes believed the trace levels of atrazine were stimulating an enzyme called aromatase, which **converts the male hormone testosterone into the female hormone estrogen.**

These findings were explosive. They suggested that atrazine, which was being used extensively in agriculture, could potentially have similar effects on human reproductive health. Hayes' research ignited a firestorm of controversy, pitting him against Syngenta and sparking a broader debate about the safety of atrazine.

In 2003, the European Union announced a ban of atrazine. Yet today, US farmers still use over seventy five million pounds of atrazine a year. Nearly 75% of US corn and sorghum land is treated with atrazine. It can also be applied on golf courses and residential lawns. The Environmental Protection Agency lowered its "level of concern" to 50 parts per billion; however, Hayes work showed that atrazine was biologically active at just 0.1 part per billion. In other words, he believes it should be off the market entirely.

Could this be linked to the dramatic decrease in measured testosterone levels in men?

POPs are persistent organic pollutants. These are toxic chemicals that are resistant to breaking down:

"Due to the slow rates of metabolism and elimination, once absorbed, POPs can persist in the body for 30 years or longer and can build up with time to toxic concentrations (Yu et al., 2011; Gallo et al., 2011). This bioaccumulation of POPs with time over many years accounts for the delayed onset of disease following initial exposure."

In other words, these chemicals accumulate over time and negatively impact our mental and physical health in every category that matters.

According to Dr. Joe Pizzorno, "Toxins are the primary cause of disease in our society".

"Toxins damage every aspect of our physiological function and play a role in virtually all diseases...This can't be dismissed as due to lifestyle, nutritional deficiencies, or the increasing age of the population".

I don't write this to fearmonger. I write this to ground you in reality, so you can face reality and work with the situation you find yourself in.

It's fascinating for me to look back at all the health practices I was trying to heal myself. I remember meditating in my New York City apartment one evening, and how incredibly difficult it was to not have automatic negative thoughts. Having gone through the protocols to decrease my toxic load since then (primarily based around sweat!), this challenge has simply disappeared. I believe my toxic load, from living on earth for 20+ years, was directly causing the psychological symptoms I was experiencing, and once the toxins were removed, they disappeared. I'm passionate about this information because I know the order of operations is important. Just like baking a cake, there are steps to doing it right. Learning about toxins and the impact they have, and then the steps to minimize your toxic burden, is something I believe every human being should be aware of so they can act on that information.

These studies and others like them highlight the importance of understanding not only the individual effects of specific toxins but also the cumulative and synergistic effects of multiple toxins. While it's impossible to completely eliminate our exposure to toxins, we can take steps to minimize our exposure and support our body's natural detoxification processes, as we'll discuss in later chapters.

It also begs the question:

What are toxins costing you?

Perhaps their impact is under the radar, just slightly lowering your energy levels throughout the day. Or you experience pangs of anxiety that don't have any logical explanation. Maybe you have a hunch that your mental clarity and joy for life aren't where they used to be when you were a kid. As I conducted my research, one of the most startling comments I heard from a leading toxin expert hit home: "the greatest indicator of your toxic load is your age." When you see young children with clear eyes, skin, and boundless energy, could it partially be due to a lower toxic load?

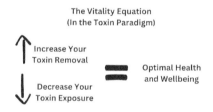

In a world where we are bombarded by countless chemicals, pollutants, and environmental stressors, it is time to recognize the profound impact of toxin exposure on our physical and mental well-being.

Every day, we encounter an invisible army of toxins – in the air we breathe, the food we eat, the products we use, and the environments we inhabit. Yet, it is all too easy to dismiss their significance, to overlook their role in the intricate web of health challenges we face. It is time to open our eyes and acknowledge the staggering impact of toxin exposure on the global stage.

This probably sounds overdramatic, but I hope the facts and statistics shared allow you to see the likelihood of this conclusion.

The steady rise in cancer rates, the alarming surge in neurological disorders, the epidemic of obesity and metabolic dysfunction, the widespread prevalence of autoimmune conditions, and the alarming increase in mental health disorders. These are not isolated incidents; they are interconnected threads woven together by the pervasive presence of toxins in our lives. It is a silent crisis, often underestimated and overshadowed by other factors, but its influence is now undeniable to many.

From the heavy metals lurking in our water and food supply to the persistent organic pollutants lingering in our homes, workplaces, and even our bodies, toxins insidiously penetrate our systems, disrupting delicate biological processes and compromising our resilience. They infiltrate our cells, disrupt our hormones, impair our immune systems, and even alter the very expression of our genes. The consequences are far-reaching, manifesting as chronic inflammation, oxidative stress, mitochondrial dysfunction, and imbalances in neurotransmitters – the building blocks of our physical and mental well-being.

As I dove into the literature around chemicals, a name kept coming up: Harold Zeliger. His research into chemical toxicology is vast. With 51 technical publications and hundreds of investigations into toxic chemical exposures, environmental spills, and chemical fires, this was a man I needed to meet.

He had a reserved, calm demeanor, even as he spoke about dark realities. I began by asking him how he'd describe toxins to the average person. "The word toxin is scary to a lot of people. But in effect, it's a substance that doesn't belong in your body that your body can't deal with eliminating."

"Chemicals are natural in the environment. Oxygen is a chemical," he went on. "We rely on sugar and proteins. The man made chemicals were not designed to be incorporated into the human body. They were designed to make plastics, fertilizers, pesticides, and they get into our body not because we want them to, but because they're present in the water supply, the air we breathe, all of our environment."

"Things like formaldehyde have very short half lives and dissipate in hours in the body. When you get into things like pesticides, PCBs, fluorocarbon compounds that are used in a variety of applications, when they are absorbed, **can last for decades**. Things like PCBs can last for 30, 40 years or more. What makes them particularly nasty is they get absorbed by the fat in your body because they're lipophilic (fat loving). They then very slowly release into your bloodstream. Most of them are carcinogens and have all kinds of horrible other effects."

I took a few things away from our conversation. Namely, that manmade chemicals are everywhere, they wreak havoc on the body, and many toxins can stay in the body for decades, quietly causing problems.

Katie Kaps is a co-founder of Higher Dose, a wellness company that helped pioneer the use of infrared saunas in NYC. As someone who has worked with hundreds of clients to heal, both in person and through Higher Dose's online business, she put it plainly: "Detoxing is no longer a nice to have thing. Everybody in this country needs to be detoxing on a regular basis."

Toxin exposure should not be a minor inconvenience or a topic relegated to niche discussions. It is a pressing global issue that demands our attention. It is an underlying force driving the epidemic of chronic diseases, the increasing burden on healthcare systems, and the untold suffering experienced by millions around the world.

—-

Solomon Asch

In a famous social psychology experiment conducted by Solomon Asch in the 1950s, participants were shown a series of lines and asked to identify which line from a set of comparison lines matched a standard line in length. However, the catch was that the majority of the participants were actors, instructed to give intentionally incorrect answers.

The true focus of the study was to observe how the genuine participants, who were unaware of the actors' instructions, would respond when faced with obvious misinformation. Astonishingly, when the confederates consistently provided incorrect answers, approximately 75% of the genuine participants conformed and also gave the wrong response at least once, despite their initial correct judgments.

This study highlights the profound influence of the herd instinct on individual behavior. Even when people knew the correct answer, they often felt compelled to align with the majority, fearing social disapproval or doubt in their own judgment. The need to conform and be accepted by the group often overshadows our own perceptions and undermines our ability to assert independent thinking.

This herd instinct can be seen across all areas of life. Taking even simple steps to protect yourself from toxins, one immediately feels that herd instinct pressure. It's not "mainstream" to put a shower filter on your shower head, or have multiple air filters in your house. It's not "mainstream" to decline the tap water at the restaurant. It's not "mainstream" to check the materials in your clothes that you wear everyday. Unfortunately, chronic disease and mental health challenges have become mainstream. Pick your poison! (no pun intended). I made a simple equation to show the framework for thinking about toxins above. In the following chapters, we're going to focus on sweat and its impact on the top half of the toxin equation: increasing your ability to get rid of toxins in your body.

Key Takeaways

Pervasive Presence of Toxins: From the moment of birth, individuals are exposed to a significant number of chemicals and toxins present in the environment, highlighting the urgent need for awareness and action.

Health Implications: Toxins have been linked to a wide range of health issues including cancer, mental health disorders, and chronic diseases. They not only affect physical health but also have a profound impact on mental and neurological health, potentially causing conditions like depression, anxiety, and developmental disabilities in children.

Toxins and Chronic Diseases: The chapter suggests a strong correlation between the rise in various chronic diseases and increased exposure to environmental toxins over the past few decades.

Sources and Types of Toxins: Toxins infiltrate our lives through various channels including industrial activities, agricultural products, household items, personal care products, food additives, and consumer products. These toxins can have adverse effects on different aspects of health, including reproductive and neurological health.

Empowerment through Knowledge: The chapter encourages readers to arm themselves with knowledge about the different categories of toxins and their impacts, fostering informed decisions about health and well-being.

5

The Science of Sweat and Detoxification

It's 2010. Stephen Genuis is a clinical researcher at the University of Alberta who has spent years studying the effects of environmental toxins on the human body. He's been fascinated by the idea that sweat could be a powerful tool in eliminating harmful substances from the body, but he doesn't have proof. But hey, he's a researcher. So he decides to put his theory to the test. He painstakingly sets up a precise, elaborate experiment.The results blow him away.

Genuis recruits 20 participants who have a history of high environmental toxin exposure and measures their blood levels of toxins before and after a sauna session.These toxins included heavy metals like lead, mercury, and cadmium, as well as environmental chemicals like polychlorinated biphenyls (PCBs) and bisphenol A (BPA).

He also collects samples of the sweat they produce during the session.

As the participants sit in the sauna, beads of sweat start to form on their foreheads and trickle down their backs. A true sauna session means profuse sweating. Multiple towels drenched in sweat by the end.

After the sauna session, Genuis analyzes the sweat samples and finds something remarkable in them: substances that weren't even detectable in the participants' blood. The substances were so toxic that the body had to store them away because it didn't know how to get rid of them until a profuse sweating session in the sauna. The discovery? Sweat is a method for detoxification.

This groundbreaking study is just one example of the growing body of research supporting the role of sweat in detoxification. The sweat produced by our sebaceous sweat glands, it turns out, is one of the body's primary methods for eliminating toxins and other harmful substances. The ancients knew this, and now science is catching up to our collective intuition. Not only that, it appears more important than ever with the introduction of manmade chemicals that our body has trouble detoxifying. Sweat is the way.

When most people think of detoxification, they think of the liver and kidneys.

While the role of sweat in detoxification is not universally accepted in the mainstream medical community, the pioneering research conducted by Dr. Genuis and others provides compelling evidence that sweat is indeed a significant pathway for the elimination of toxins from the body.

These studies have shown that a variety of toxins, including heavy metals and environmental chemicals, can be found in sweat, often at higher concentrations than in blood or urine. This suggests that the body may use sweating as a mechanism to rid itself of these harmful substances, particularly when other detoxification pathways are overwhelmed or less effective. Let's not forget: Genuis' study took place in 2010, just 13 years prior to the writing of this book.

Just back in 2020, a scientific study investigated the excretion of certain heavy metals through sweat. The study was conducted by researchers from the Department of Environmental and Occupational Health, College of Public Health, University of Arkansas for Medical Sciences, Little Rock, AR, USA.

The study was conducted with the aim of understanding the role of sweating in the excretion of heavy metals from the body. The researchers specifically looked at Nickel (Ni), Lead (Pb), Copper (Cu), Arsenic (As), and Mercury (Hg). The study involved two different sweating conditions - exercise-induced sweating and sauna-induced sweating.

The results of the study showed that all five metals were excreted in sweat under both conditions. Heavy metals, particularly arsenic and mercury, are known to be extremely toxic to human beings.

Moreover, the practice of induced sweating, such as through sauna sessions, has been used for centuries in various cultures around the world for its perceived health benefits. This traditional wisdom, coupled with emerging scientific evidence, points to the potential of sweating as a natural and effective method for detoxification.

While more research is certainly needed to fully understand the mechanisms and optimize the use of sweating for detoxification, the existing evidence supports the idea that sweating is an important, and perhaps underappreciated, aspect of our body's detoxification system. And, as we'll see in future chapters, the case studies of extremely toxic people being cured through sweat leaves me certain that this is true.

As we continue to grapple with the increasing burden of environmental toxins, harnessing the power of sweat has proven itself to be an invaluable strategy for enhancing health and preventing disease.

But how exactly does this process work? How does sweat help to detoxify the body?

As discussed earlier, sweating is the body's way of regulating internal temperature. When we sweat, toxins are excreted from the body through the sweat glands. This process is known as "dermal elimination," and it occurs when toxins are released from the blood into the sweat glands and then eliminated through the skin. This is an important mechanism for detoxifying the body, as it allows harmful substances to be removed from the body without putting additional strain on the liver and kidneys.

Sweating also stimulates the lymphatic system, which is responsible for removing waste and toxins from the body. The lymphatic system is a network of vessels and nodes that work together to circulate lymph, a fluid that carries white blood cells and other immune system cells. When we sweat, the movement of the lymphatic fluid is increased, which helps to remove toxins and waste products from the body.

Another way that sweating helps to detoxify the body is by increasing circulation. When we sweat, blood vessels near the surface of the skin dilate, allowing more blood to flow to the skin. This increased blood flow brings more oxygen and nutrients to the skin, and it also helps to remove toxins and waste products from the body.

Sweating is a crucial mechanism for detoxifying the body. By allowing harmful substances to be eliminated through the skin, stimulating the lymphatic system, increasing circulation, and producing endorphins, sweating plays a vital role in maintaining optimal health and wellbeing.

6

Sauna

A Sauna Adventure

In May of 2023, as the world was slowly recovering from the grip of the pandemic, I found myself seeking a sense of stability and belonging. New York City, with its vibrant energy and endless possibilities, called me back. With the help of my brother, a seasoned real estate agent, I discovered a gem of an apartment nestled in the heart of Williamsburg, Brooklyn. It was a place I could finally call home, at least for the foreseeable future.

But as I settled into my new urban surroundings, I couldn't ignore the reality of city living and the constant exposure to environmental toxins. The bustling streets, pollutants in the air, and the daily hustle and bustle of city life made me acutely aware of the need to prioritize my health and well-being.

That's when the idea struck me— I needed a sauna. A sanctuary where I could escape the chaos and cleanse my body and mind. And what better way to achieve that... than with a sauna?

Here's the thing: when you mention buying a sauna for your NYC apartment to friends and family, you are met with a mix of curious looks, raised eyebrows, and even some playful teasing. Getting a sauna in your apartment is still considered unorthodox, even in a city known for its progressive mindset.

I realized that people's reactions were often fueled by misconceptions and preconceived notions. To some, it sounded like I was jumping on the latest wellness fad, chasing after the next trendy obsession. Others thought I was becoming a zealous advocate for sweating, as if I had discovered some secret society of sauna enthusiasts.

In truth, my decision to bring a sauna into my home was grounded in a genuine desire to prioritize my health and well-being. I had come to believe that regular sweating was the single most important step in having maximum vitality and preventing disease. If that's what you believe, wouldn't getting a sauna be a no brainer?

Fortunately, my brother saw it the same way. Since he happened to be living in the same neighborhood, we decided we'd split the cost and get a sauna together.

I meticulously scribbled down a list of potential options to buy, determined to find the perfect sauna that would seamlessly fit within the cozy confines of my apartment. It had to be more than just a space-saving solution; it needed to embody the principles of low toxicity and minimal electromagnetic fields (EMF). Little did I know that this search would uncover an inconvenient truth

– that finding a sauna brand that met these stringent criteria was no easy feat.

In my pursuit of finding a sauna that met my stringent criteria, I miraculously stumbled upon a Facebook group that proved to be an invaluable resource. With an impressive membership of 50,000 passionate sauna enthusiasts, this vibrant online community provided me with a wealth of knowledge and insights into the world of saunas.

In addition to its unrivaled insights into saunas, the Facebook group offered something even more powerful: a community united by a shared journey through health challenges. Members candidly shared their personal stories, opening up about the trials and tribulations they faced in their quest for improved well-being. Their experiences spanned a wide spectrum, from chronic pain and fatigue to autoimmune disorders and environmental sensitivities. Some dealt with cancer, diabetes, Alzheimer's, you name it. The number of testimonials for how sauna use was a major contributor to their healing was remarkable.

What struck me was the resilience and determination in this group of individuals. They had encountered countless roadblocks and had struggled to find effective solutions to their health challenges. Their collective wisdom had been forged through countless hours of research, consultations with experts, and trial and error with various wellness practices. As a result, they possessed a level of discernment and scrutiny that went far beyond the average person seeking wellness information.

It was inspiring to witness the relentless pursuit of truth within this community. Their shared experiences had fostered a deep commitment to scrutinizing claims, digging beneath surface-level marketing, and seeking evidence-based approaches. Armed with their hard-earned knowledge and insights, they were determined to navigate the foggy landscape of the sauna industry, uncovering the truth and forging a path toward genuine wellness.

As I delved deeper into the group's discussions, I discovered a remarkable dedication to testing the truth behind sauna claims. Members took it upon themselves to separate fact from fiction, using tools like VOC reports and EMF meters to conduct meticulous home tests. They would use VOC reports, which detect volatile organic compounds (toxins), and EMF meters, which measure electromagnetic fields, and upload photos of their tests.

Their work and journey resonated with me. It reminded me that when faced with health challenges, we become our own advocates, diligently seeking answers and refusing to accept superficial solutions. The health struggles they shared made me acutely aware of the importance of careful discernment, the value of evidence-based information, and the need for transparency in the pursuit of wellness.

Fortunately during this research, I came across a Canadian brand that passed the group's tests. And, lucky for me, they had a unique product on the market that was sized to fit an apartment with ease.

I called up the owner, had a conversation to get some questions answered, and decided this was a smart decision. A couple clicks on my laptop later, and I began to eagerly await the arrival of my apartment sauna.

Having a sauna at home has been wonderful. But even if you can't get one at home, there are many alternatives we'll get into in this chapter. One thing is for certain: sauna is a game changer. Below, let's find out why.

Sauna Therapy

We've talked about sweat lodges and saunas in the context of their use throughout history. Let's get specific about sauna therapy, what it is, and how it can transform your health.

As I mentioned in the introduction, saunas changed my life. This book was largely a by-product of me figuring out why. My hope in this chapter is that your entire view of saunas changes, and you begin to see what I, along with so many others in the health community, see: sauna is one of the most powerful, profound activities you can do for your health, wellbeing, and disease prevention.

What is sauna therapy?

Sauna therapy is a health practice that involves exposing the body to high temperatures for the purpose of improving overall health and wellbeing. There are several different types of saunas, the main ones being (from least effective to most effective for health improvement and detoxification) steam rooms, dry saunas, and infrared saunas.

Steam rooms are typically made of ceramic or stone and are heated by boiling water, which produces steam. The high humidity in steam rooms can be beneficial for respiratory health, as it can help to clear the sinuses and improve breathing. Additionally, the steam can help to open up pores in the skin, allowing for more effective sweating and toxin elimination. Here's the concern with steam rooms: the water in the air. You're breathing in the water vapors and have no idea what that water contains (and if you Google search "EWG tap water" reports for the water in your area, you might not like what you find). Additionally, there's an argument that the water in the steam room's air inhibits your ability to sweat. Steam rooms can also be expected to be prone to mold, given the humid conditions. This is not an issue really with dry or infrared saunas.

Dry saunas, also known as traditional saunas, are typically made of wood and are heated by a wood-burning stove or an electric heater. Unlike steam rooms, dry saunas have low humidity, which can make the heat feel more intense. Dry saunas are often preferred by individuals who want a more intense heat experience. Expect to sweat profusely in a dry sauna (which is what you want). I've spoken with friends who love dry saunas because they feel like a test of endurance and mental fortitude.

Infrared saunas use infrared heaters to emit infrared light, which heats the body directly instead of heating the air around it. As a result, the experience isn't as intense as a dry sauna;

however, it causes the body to sweat even more. Some fascinating studies have come out supporting the idea that infrared saunas promote even more profound detoxification than dry saunas.

One study published in the Journal of Human Kinetics compared the effects of dry sauna and infrared sauna use on athletes. The study found that while both types of sauna use were effective at promoting sweating and increasing heart rate, infrared saunas were more effective at promoting the elimination of toxins from the body.

A potential reason for the increased effectiveness of infrared saunas is that they can penetrate deeper into the skin than dry saunas, which may allow them to target and release toxins that are stored deeper in the body.

Sauna use is certainly growing in popularity in the United States, but is certainly not a part of the culture yet. A 2015 study showed that approximately 10% of Americans had used a sauna at least once in the past year. There aren't any more recent studies or analysis of regular sauna users (at least 3 times a week), but we can assume this number is still quite low.

Regardless of the type of sauna used, the general principle behind sauna therapy is the same. Exposure to high temperatures causes the body to produce sweat, which is a natural mechanism for eliminating toxins and waste products from the body. Additionally, sauna therapy can help to improve cardiovascular health, reduce inflammation, enhance immune function, and improve athletic performance.

So there you have it. The three main types of saunas. Let's take a deeper dive into the miraculous health benefits of sauna outside of detoxification.

- **Mood Enhancement**: Sauna use can lead to the release of endorphins, the body's natural feel-good chemicals, which can improve mood and reduce anxiety. This is due to the activation of the body's stress response pathways, which can lead to a net resilience effect, helping individuals handle stress better. People who sauna regularly often report better, more stable moods throughout the day.
- **Cardiovascular Health**: Regular sauna use can have a positive impact on cardiovascular health. Studies have shown that men who use a sauna 2-3 times a week are 27% less likely to die from any cardiovascular-related disease compared to men who use a sauna once a week. Those who use a sauna 4-7 times a week are 50% less likely to die from cardiovascular-related diseases. The duration of sauna use also matters, with longer sessions providing more benefits. In other words, the more you sauna, the better for your heart health! (within reason)
- **Longevity**: Sauna use can also affect overall longevity. Regular sauna users were found to have a 24% lower all-cause mortality rate than those who used the sauna less frequently. This includes a lower likelihood of dying from cancer, cardiovascular disease, neurodegenerative disease, and respiratory diseases.
- **Heat Shock Proteins**: Sauna use leads to the activation of

heat shock proteins (HSPs), which play a crucial role in maintaining the three-dimensional structure of proteins in the body. This is important because damaged proteins can lead to diseases such as Alzheimer's and cardiovascular disease. HSPs can repair damaged proteins, preventing them from forming harmful aggregates.

- **Activation of FOXO3**: Sauna use can activate the FOXO3 gene, a master regulator of many other genes related to stress resistance. This can help the body handle the normal damage it experiences every day, as well as damage from external factors like air pollution and carcinogens.

- **Exercise Mimicry**: Sauna use can mimic the effects of moderate-intensity physical exercise. It increases heart rate, expands plasma volume, and increases blood flow to the heart, which lowers cardiovascular strain and improves endothelial cell function and left ventricular function. Those who start to sauna are often worried by this increased heart rate while simply sitting there; however, there's no reason for alarm. In fact, it's like your getting a light workout in without the normal effort!

In summary, you feel better, live better, and live longer with regular sauna use.

When I bring up saunas to someone who doesn't use them regularly, they often bring up Finland.

The use of saunas in Finland dates back thousands of years, and the sauna remains an important part of Finnish culture to this day. Saunas are found in homes, apartments, and public buildings throughout the country, and are used for relaxation, socializing, and for their health benefits.

The Kalevala, a Finnish epic poem, includes references to the use of saunas. The Kalevala, which was compiled and edited by Elias Lönnrot in the 19th century, is based on traditional Finnish oral poetry and mythology. In the poem, the sauna is described as a sacred space, and the act of sweating in the sauna is seen as a way to purify both the body and the mind.

In one famous passage of the Kalevala, the hero Väinämöinen creates the first sauna by carving it from a tree, heating it with stones, and filling it with steam. This act is said to have brought health and vitality to the people of Finland, and to have established the sauna as a cornerstone of Finnish culture.

When I started writing this book, my goal was to transplant some hard-won truths through years of searching through the healthcare system. My hope is to get you to believe a few things:

- Sweating profusely is something that cultures around the world have intuitively understood the value in for millennia
- We are exposed to a ludicrous amount of toxins that are largely responsible for many of the health issues facing the modern human
- Sweating is an absolutely critical way to reduce the toxic load in your body

- Saunas are the easiest, most effective method for sweating
- Infrared saunas are the most effective of the sauna options for detoxification

If you care about your health and vitality, sweating is for you.

And, if you understand why and how sauna is a game changer for your health, you're probably wondering how to get started.

How to Find a Sauna Near You

Having regular access to a sauna is key to realizing its many health benefits. If you don't have a personal sauna at home, don't worry! There are many local saunas available, and here's how you can find them.

Google Maps Search: The quickest way to find a sauna in your area is in Google Maps.

1. Google Maps Search: Pull up maps.google.com, type in your area, and then search for "sauna". This should bring up a list of nearby establishments that offer sauna facilities, including spas, health clubs, and gyms.
2. Call and Ask: Once you have a list of potential places, call each of them to gather more information. Ask about the type of sauna they offer. As we've mentioned before, infrared saunas are ideal, dry saunas are still great, and steam saunas are best avoided.
3. Amenities: Don't forget to ask about additional facilities like showers, as you'll want to rinse off after a sauna session.
4. Packages and Discounts: Many places offer discounted

rates for a package of sessions. Make sure to inquire about these deals as it can make your sauna experience more affordable.

5. Gyms: Don't overlook gyms! Many fitness centers have dry saunas as part of their amenities, and this could be a cost-effective option if you're already a member.
6. Location: Finally, consider the location of the sauna in relation to your home or workplace. Regular sauna use will be easier if it's conveniently located.

By following these steps, you should be able to find a suitable sauna near you that fits your needs and budget. Depending on where you are, there will be a higher or lower quantity of options available in your area.

If you're new to sauna, this is a great way to get acclimated to the process before purchasing a sauna. If you live in a city and don't have much room in your apartment, this is also an excellent alternative.

The Sauna Session: How To Sauna

Embarking on your sauna journey is an incredible step towards optimal wellness, but it's important to understand how to use the sauna correctly. A sauna session isn't a race, but a gradual process of acclimatization and enjoyment. Here's a beginner's guide to a sauna session:

Duration: Start with sessions of around 25 minutes, preferably in a dry or infrared sauna. As your body adapts to the heat, you can gradually increase the duration. Remember, slow and steady wins the race. Eventually, you'll feel comfortable spending up to an hour in the sauna. Ideally, you want to have profuse sweating for at least 20 minutes.

Listen to Your Body: If you feel lightheaded, excessively hot, or otherwise uncomfortable, don't hesitate to step out of the sauna and take a break. Hydrate with water, ideally with added electrolytes, to replenish lost fluids.

Acclimatization: Over time, your tolerance to heat will improve. You may find that after about 10 sessions, spending an hour in the sauna with 2-3 quick breaks for water feels comfortable.

Relaxation and Mindfulness: Use your time in the sauna to meditate, relax, or engage in quiet conversation if you're sharing the space with others.

Attire and Items: Enter the sauna with cotton towels. Avoid bringing plastic items, such as water bottles, into the sauna to prevent the release of toxins in the heated environment. It is a hotbox, after all, and you don't want to be breathing that in. Phones can be alright in dry and infrared saunas, but be cautious of overheating. Don't wear clothing or shoes either! Cotton towels work, but most clothing will leach toxins into the air.

Sweating: Once you start sweating profusely, wipe your body down regularly with your towels to aid the sweating process. You can bring extra towels, or step out to grab another when one becomes saturated. I typically wear a towel around my waist if I'm in a public sauna, or go in naked with a couple towels if I'm in a private sauna.

Hydration and Replenishment: After your sauna session, it's crucial to rehydrate and replenish your body's electrolytes. You lose essential minerals through sweating, so consider a balanced electrolyte drink to replenish your system. Real spring water (not the plastic bottled brands, which often use reservoirs) has a decent amount of minerals like Potassium and Magnesium in it. When looking for this in a store, glass bottled spring water tends to be your best bet. If you look at the bottle, you also want to see that it was sourced from a single location.

Remember, everyone's body responds differently to heat, so listen to your body and adjust your sauna practice accordingly. By following these principles, you'll be able to enjoy your sauna sessions safely and effectively, reaping the many health benefits they offer.

Saunas: What to Avoid (whether buying one or using a public one)

The sauna industry is largely unregulated, which can lead to companies cutting corners on construction and safety. As you familiarize yourself with the world of saunas, it's important to be aware of potential pitfalls. Here are the main things to watch out for:

Toxic Materials: When using a sauna, you're essentially in a heated box, breathing in the air within. If the sauna materials are toxic, these toxins can migrate with the heat, contaminating the air you breathe. Many saunas use materials like bamboo or flame retardants that, when heated, can release harmful substances into the air. This is counterproductive to the detoxification process you're trying to promote with sauna use. Always ensure that the sauna you're using is built from non-toxic, natural materials.

However, it's worth noting that public saunas, which are often on for extended periods each day, can off-gas a large percentage of any toxic materials that are part of their construction. This is similar to a new car: initially, it has a strong smell due to the chemicals in the seats and other parts. Over time, as the car is exposed to sunlight and heat, it off-gases most of these chemicals, reducing the smell and potential toxicity. Public saunas might be on for 12 hours a day, so the risk is lower than if you buy one and have it on for maybe 30 minutes to a couple hours a day.

Despite this, it's always better to err on the side of caution. Try and ensure that the sauna you're using is built from non-toxic materials.

Electromagnetic Fields (EMFs): EMFs are invisible areas of energy, often referred to as radiation, associated with the use of electrical power and various forms of natural and man-made lighting. While low-level EMFs are generally viewed as harmless, higher levels can cause health issues over time, such as headaches, fatigue, and sleep disturbances. Unfortunately, some sauna companies may provide misleading information about their EMF levels. Always look for third-party testing and certification of EMF levels when choosing a sauna.

Up next, we're going to talk about arguably the most powerful detoxification protocol in the world that has saunas at the center of its practice. Saunas alone are a singularly powerful tool in maximizing our health and preventing disease. This protocol we're about to explore takes it to another level.

7

The Niacin Sauna Protocol

I believe things become obvious when taken to extremes. When we have extreme issues, we can see more clearly into how potential solutions could work. So let's look at an example of one of history's most intense examples of toxicity: September 11th.

Many first responders and volunteers who worked at Ground Zero in the weeks and months following the attacks were exposed to a wide range of toxins and pollutants, including asbestos, benzene, and other hazardous chemicals. Clouds of debris filled the air across downtown Manhattan, and anyone unfortunate or courageous enough to be in the area was inhaling these toxins.

Toxic substances included asbestos, radionuclides, benzene, dioxins, polychlorinated biphenyls (PCBs), fiberglass, mercury, lead, and silicon. These agents are associated with cancer as well as severe lung pathology, neurological and cardiovascular disease, and a myriad of immune dysfunctions. All of which became prominent in 9/11 survivors.

First responders and emergency workers involved in the rescue and cleanup efforts after the September 11th attack were subjected to hazardous chemicals and their breakdown products at an unprecedented level. They were exposed to these toxins for over eight and a half months, working long hours every day without proper Personal Protective Equipment (PPE) to protect them from inhalation, ingestion, or dermal exposure.

Unfortunately, the PPE provided was generally ineffective, leaving them vulnerable to the toxins in the air. This exposure resulted in devastating health consequences for many of the workers, who suffered from a range of symptoms and diseases due to the toxic exposure.

The numbers impacted by this were in the tens of thousands.

Exposure to the toxins and pollutants at Ground Zero resulted in many first responders and volunteers reporting a wide range of symptoms that were both physically and mentally debilitating.

Respiratory problems were a common issue, with many individuals suffering from persistent coughing, wheezing, and shortness of breath. Some even developed chronic respiratory conditions such as asthma, bronchitis, and chronic obstructive pulmonary disease (COPD). Eye irritation and skin rashes were also frequently reported.

In the aftermath of 9/11 and with thousands suffering, something had to be done.

A group of firefighters and union representatives believed that a program was necessary to help the rescue workers detoxify from their massive exposures. They reached out to the Foundation for Advancements in Science and Education (FASE) to help them create a detoxification plan for those who had been exposed to the toxins.

After reviewing all the options for a solution, they decided on the Niacin Sauna Protocol.

85

In September 2002, an independent facility was established in Lower Manhattan, funded by private donations, to provide free therapy to those affected by toxic exposure at the WTC site.

In the initial study, 500 people were able to go through the program. Ultimately, close to 5,000 people went through it. While this was not a controlled study and was in the context of outcome monitoring, the results are still nothing short of miraculous.

- All 500 participants reported improvement in subjective symptoms.
- All participants reported improved perception of health.
- Health History and Symptom Survey (selected questions) found considerable reductions in days of work missed on the start of the detoxification program, leading to reduced concerns about forced retirement.
- Due to symptom improvement, 84% of those participants requiring medications to manage symptoms related to WTC exposure were able to discontinue their use.
- Over half the participants required multiple pulmonary medications on entry to achieve near-normal pulmonary functions (measured as FVC & FEV1). On completion of detoxification, 72% of these individuals were free of pulmonary medication yet had improved pulmonary function tests (data not shown).
- There was a statistically significant improvement in thyroid function tests.
- There was a statistically significant improvement in Choice Reaction Time (CRT) and Intelligence Quotient (IQ), sug-

gestive of improvement in cognitive function.

- Statistically significant improvement in Postural Sway Test that indicated improvement in vestibular function.
- The "Postural Sway Test" is a test that measures how much a person's body sways while standing still with their eyes closed. It is used to evaluate a person's balance and stability. A decrease in postural sway means that a person is more stable and balanced. Therefore, a statistically significant improvement in the Postural Sway Test suggests that the participants' vestibular function improved. The vestibular system is responsible for providing sensory information about motion, equilibrium, and spatial orientation, and it plays an important role in balance and stability.

Imagine 500 people suffering from debilitating symptoms. Missing work. Cognitive function diminished. Chronic fatigue. Difficult with balance and stability. They each tried multiple approaches to heal, none of which made a significant difference. They then all went through this protocol, and every single one of them healed. This is simply too significant to ignore!

For people outside of this program, many believed the symptoms would decrease over time. Tragically, this was usually not the case. In fact, many who didn't initially report issues began falling ill years after the exposures.

As I read through the case studies and firsthand experiences of those who went through the program, I reached out to Anne-Marie Principe.

Anne-Marie's Story

"It was a toxic cloud of which no one had ever experienced before. It was the most god awful mix of chemicals."

Over a Zoom call one afternoon, I was greeted by Anne-Marie Principe. With long blonde hair and glowing skin, she seemed like the picture of health. But her story was far different, and I can tell by her warrior spirit that she had been through hell and came out the other side. As I introduced myself, I could feel her excitement at the opportunity to share her story.

"I was downtown on September 11th, in the street. First plane hit, we didn't know what it was because we didn't see it. We remained in the street, don't ask me why. When we felt the ground shake, we knew those buildings were coming down. I was fortunate, I didn't get hit with anything, but I was completely covered in dust. Immediately, I was fighting to breathe and didn't feel well. I went to the doctor probably 9 hours after the towers were hit. It just got progressively worse as the months went by."

Anne-Marie explained how she went to multiple specialists, before ultimately being referred to the New York Rescue Workers Detoxification Project.

"I was blown up from the steroids my pulmonologist had given me. On top of the steroids, I was taking multiple inhalers just so I could breathe. When I went into the detox, all of us were so sick. Initially, our group was 15 of us. The protocol was you took your niacin, got on a bicycle or treadmill if you could, and then you went into the sauna and sweated it out. I can tell you that what came out of my skin and system was insane. I had plugs of mucus coming up that were grey, and I had ash coming out of my skin. It took me a good month and a half before I started to feel different. And within 3 months, I was off every medication but my rescue inhaler. I wasn't yellow, I dropped about 30 pounds. For the first time, I could breathe again without equipment. I could be a mom to my daughter again. It just made a world of difference."

Anne-Marie described how everyone in her group had similarly profound results:

"You could see that all of these sick, weak, emotionally challenged people were all of a sudden talking, joking around... you could see color return to their faces. It was total, utter transformation."

Now that we've covered the results of the Niacin Sauna Protocol, let's talk about the protocol itself.

What is the Niacin Sauna Protocol?

At a high level, the niacin sauna protocol involves 5 steps:

1. Take a flushing dose of niacin. People start with 100mg and work their way up.

2. Wait two and a half hours for lipolysis to begin (we'll cover what lipolysis is shortly)

3. After two and a half hours, engage in 30 minutes of exercise

to help facilitate the release of toxins from your fat cells.

4. After exercising, sit in an infrared sauna for around an hour and 15 minutes to sweat out the toxins.

5. After leaving the sauna, take any necessary binders/supplements and shower to wash off any remaining toxins that have been released from your body.

Let's look closer at each of these parts so we can understand how they work together to create such magical results.

Take a flushing dose of niacin

First, what the heck is niacin? Niacin is actually vitamin B3.

In the 1960s, Abram Hoffer and Dr. Humphrey Osmond were psychiatrists who believed that many mental illnesses could be treated by using vitamins and other natural substances. They found that high doses of vitamin B3, or niacin, were particularly effective for treating schizophrenia.

One of their patients was Dr. Andrew Saul Kaufman, a psychiatrist himself. Kaufman had been suffering from depression and decided to try the niacin therapy that Hoffer and Osmond had been experimenting with.

To Kaufman's surprise, not only did the niacin help lift his depression, but it also improved his overall health. Kaufman became an enthusiastic advocate of niacin therapy and began experimenting with it in his own practice.

In the 1970s, Kaufman began to develop his own version of the niacin protocol for detoxification. He found that high doses of niacin, combined with extended periods of time in a sauna, could help remove toxins from the body and improve overall health.

Niacin, also known as vitamin B3, is a water-soluble vitamin that plays an important role in various bodily functions, including energy production, DNA repair, and cell signaling. It is also essential for maintaining healthy skin, nerves, and digestion.

Niacin is found in many different foods, including meats, fish, eggs, milk, and green vegetables. It can also be synthesized by the body from the amino acid tryptophan, which is found in high-protein foods like meat, fish, and nuts.

One of the unique characteristics of niacin is its ability to cause a flushing sensation in the body when taken in large doses. This flushing is caused by the expansion of blood vessels near the skin's surface, which can cause a tingling or warm sensation, redness, and occasionally itching. This flushing response is harmless and typically lasts for about 30 minutes to an hour.

In addition to its normal functions, niacin has been found to have a detoxifying effect on the body, particularly in relation to removing toxins stored in fat cells. Niacin helps to break down these fat cells, releasing the stored toxins into the bloodstream, where they can be filtered out by the liver and kidneys. This point is critically important. Remember earlier how we talked about lipophilic (fat loving) toxins? Niacin is the most effective known way for rupturing these fat cells to release the toxins. This is NOT a good idea, as it floods your body with toxins that the body was hiding away; however, combined with the sauna, it allows you to sweat out potentially years of stored toxins.

Wait two and a half hours for lipolysis to begin

Lipolysis is the process of breaking down fat molecules into smaller molecules called fatty acids and glycerol, which can then be used by the body for energy. Niacin, or vitamin B3, plays a key role in this process by activating an enzyme called lipoprotein lipase (LPL), which is responsible for breaking down the triglycerides stored in fat cells into fatty acids and glycerol.

When niacin is taken in large doses, it causes a temporary increase in blood flow, which triggers the release of fatty acids from the fat cells into the bloodstream. This process is known as flushing, and is often accompanied by a warm, tingling sensation on the skin. The released fatty acids can then be burned for energy during exercise or in the sauna, leading to increased detoxification.

When you take niacin, here's what happens:

1. Niacin is absorbed into the bloodstream.
2. Niacin is converted into NAD+ in the body.
3. NAD+ (Nicotinamide Adenine Dinucleotide) is a coenzyme found in all living cells. It plays a vital role in energy metabolism and is essential for the functioning of many enzymatic reactions within the body. NAD+ is involved in

various metabolic reactions, including the breakdown of fats.

4. The breakdown of fats (lipolysis) releases stored toxins into the body.

Niacin causes these lipophilic toxins that had been stored by the body for perhaps years to get released.

Importantly, one should time niacin with sauna use so that these circulating toxins can be properly excreted from the body through sweat.

The niacin used in the protocol is a specific form called "nicotinic acid," which is different from other forms of niacin such as niacinamide or inositol hexanicotinate. Only nicotinic acid has been shown to have the lipolysis effect that is necessary for the protocol to work effectively. Additionally, it is important to use a high-quality form of niacin that is free of contaminants and fillers, as low-quality forms can cause liver damage and other health problems.

Engage in 30 minutes of exercise to help facilitate the release of toxins from your fat cells.

The idea is to time when to get in the sauna to optimize the amount of toxins you sweat out. From my personal research and understanding of what people have found, two and half hours of waiting followed by 30 minutes of exercise is optimal.

During the niacin sauna protocol, the exercise component is

crucial because it helps to mobilize the stored toxins from the adipose tissues and into the bloodstream, where they can be more easily eliminated through sweating in the sauna. Additionally, exercise helps to increase circulation and oxygenation to the tissues, which can aid in the removal of toxins from the body.

By exercising prior to entering the sauna, the body is better prepared to handle the increased demand for energy and heat stress that will be encountered during the sauna session. This helps to ensure that the body is able to effectively detoxify and eliminate the stored toxins that have been mobilized from the adipose tissues.

Sit in an infrared sauna for an hour and 15 minutes to sweat out the toxins

Infrared saunas are the best choice, followed by dry saunas. There are caveats to this as we discussed in Chapter 6. During this hour and fifteen minutes, you're going to be sweating like crazy. The hotter the sauna, the better. Bring in multiple towels, continue to wipe yourself down. People typically do a 45 minute session, rest for 10-15 minutes, then finish with a final half hour. If you feel unwell at any point, don't hesitate to take a break outside the sauna for a couple of minutes, and stay hydrated. Just like the gym, people build their heat tolerance and capacity to stay in the sauna longer. For some, 20 minutes straight. For others, an hour straight is a breeze.

After leaving the sauna, take any necessary binders and shower

After finishing the sauna session, it is important to take a shower to rinse off any toxins that may have been released through sweat.

In addition to showering, it is recommended to take binders, such as activated charcoal, zeolite, or chlorella, to help absorb any toxins that may have been released from fat cells during the session. Binders work by binding to the toxins and preventing them from being reabsorbed into the body. The niacin has released a tremendous amount of toxins from the fat cells, and the sweat has allowed the body to eliminate much of them; however, many will still be circulating in the body. Binders grab onto them and allow you to eliminate more effectively in the bathroom.

What does this mean for you?

So now you have an overview of the protocol. You understood the use case for 9/11 survivors. But how does this relate to the average person?

While the niacin sauna protocol was initially developed to help detoxify drug addicts, it has since become a powerful tool for anyone looking to improve their overall health and well-being.

A study published in the journal Environmental Health Perspectives in 2006 analyzed the blood and urine samples of a representative sample of Americans and found that participants had an average of 91 toxins detected per person. These toxins included heavy metals, pesticides, flame retardants, and many other harmful substances.

While our bodies have natural mechanisms for removing toxins, they are often overwhelmed by the sheer volume and variety of chemicals we encounter in modern life. This can lead to a buildup of toxins in our fat cells and other tissues, which can have a wide range of negative effects on our health, including chronic diseases like cancer, neurological disorders, and autoimmune conditions. Everything from low level fatigue, lack of joy, and brain fog to depression, cancer, and Alzheimer's has been shown to have causal relations to these toxins.

As demonstrated in Genuis' experiment, a single sauna session can be shown to help release some of these toxins. The niacin sauna protocol takes this to the next level, by releasing stored toxins in the fat cells for detoxification. Niacin's magic is in its ability to open the fat cells, release the stored toxins your body couldn't get rid of, and provide the opportunity to sweat it out.

9/11 survivors are an extreme example that illustrates the power of this protocol. Does it make sense for the average person? Personally, I believe it does. Let's look at the evidence we've built up once again.

1. We're exposed to ludicrous amounts of toxins that our body has trouble detoxifying
2. They are unequivocally linked with the majority of health problems facing us
3. Our toxic burden only increases with age

When we go to step on a scale, we get an exact measurement of where we stand. If we're overweight, there's no getting around it. Unfortunately, with today's technology, we can't do a scan of our body and get precise measurements of what toxins are inside of us. If we could, I think we'd all be in for a shock. The niacin protocol is, in my opinion, a miracle solution to a modern epidemic.

When properly supervised, the risks are minimal, but the detoxification upsides are tremendous.

We're all exposed to massive amounts of toxins. This protocol is the most effective process for detoxifying years of stored toxins from the body.

While this protocol has minimal risk, it cannot be advised to be undertaken without professional support. I will be linking updated resources on my site sweatthebook.com for more information. But, again, do NOT undertake this process by yourself. Start with a sauna and sweat, as it's still an excellent place to begin.

I reached out to some people who had been through the program to gather testimonials. Below are some of those people's experiences with the protocol:

"At the end of the program, I felt like I was 30 years younger! Higher energy all throughout my work day. My doctor is amazed at the results of my blood work and all other significant health markers. He said that my body is performing like a person aged 40 or less. There is no evidence I have ever been ill at an earlier age. I will be 81 in early September and plan to work as a CEO and founder of companies until well into my 90th year of life.

Some of my superior health can be credited to mental processes I used to recover from Chron's Disease and two strokes, but I give full credit to the detox protocol which gave me a "second wind," boosted my energy level, and kept my mind clear." - Jon Taber, 81

"I had gadolinium poisoning, heavy metal poisoning, multiple chemical sensitivity, along with severe migraines (6-7 a week), unhealthy gut biome, accid reflux, low functioning gallbladder, chronic fatigue and low energy, chronic sinusitis, chronic ringing in my ears, floaters in my eyes, environmental allergies with allergy shots, asthma, endometriosis, pain from neck, back and shoulder injuries with limited mobility, muscle and joint pain, and chronic swelling in my legs from lipedema. These challenges made me feel like I was broken at the age of 35 and I had no light at the end of the tunnel. I felt like I was going to be sick for the rest of my life and never going to be able to return to work. After the protocol, My quality of life significantly increased and I was able to return to work full-time. I am now 39 and still going strong from only one detox. My plan is to complete another once I purchase my own sauna. I would encourage everyone to do this detox. " - Jessica Rivers

"From doing the protocol, my energy is through the roof, less joint pain, clearer thinking, weight loss after detox, less food intolerances, and lower blood pressure." - Melissa Hines

8

Conclusion

The concept of gyms as we know them today took root in the early 20th century, as a growing interest in physical fitness and bodybuilding swept across America.

In the late 1800s, a cultural shift began to take place as the industrial revolution transformed the American way of life. The rise of sedentary occupations and modern conveniences led to concerns about declining physical fitness and overall health. Influential figures such as Bernarr Macfadden and Charles Atlas emerged, advocating for the importance of exercise and strength training.

During the early 1900s, the popularity of gyms started to rise. These establishments, often referred to as "physical culture studios" or "athletic clubs," offered spaces for individuals to engage in various forms of exercise and physical training. These early gyms primarily catered to men and focused on weightlifting, bodybuilding, and other forms of strength training. This new concept was influenced by various factors, including advancements in exercise science, growing interest in physical aesthetics, and a desire for structured fitness programs.

In the mid-20th century, the popularity of gyms expanded further with the rise of fitness and wellness movements. The advent of modern exercise equipment, such as treadmills, stationary bikes, and weight machines, made exercise more accessible to a wider range of people. This era also saw the emergence of fitness pioneers like Jack LaLanne, who popularized the concept of regular exercise and healthy living through his television show and fitness studios.

The fitness industry continued to evolve and diversify throughout the latter half of the 20th century and into the 21st century. The introduction of aerobics and group exercise classes, the proliferation of commercial gym chains, and the emphasis on holistic wellness contributed to the widespread adoption of gym culture.

Today, gyms have become an integral part of culture, offering a variety of fitness options, specialized training programs, and wellness services. They serve as community hubs where individuals of all ages and fitness levels can pursue their health and wellness goals.

The history of gyms in America reflects a societal shift towards prioritizing physical fitness and recognizing the profound impact that regular exercise can have on overall well-being. From humble beginnings to modern fitness centers, gyms have played a crucial role in shaping our understanding of health and promoting the importance of physical activity in our daily lives.

In the wide world of health and wellness, sweat has been overlooked for far too long. Throughout this journey, we have delved deep into the many facets of sweat, unraveling its intricate mechanisms, exploring its profound impact on our well-being, and uncovering its role in detoxification and revitalization. We have discovered that sweat is not just a byproduct of physical exertion or a response to heat; it is a remarkable "detoxifier" and a powerful catalyst for optimal health.

Yet, despite its now undeniable importance, sweat hasn't gotten the credit it deserves. We may often hear a friend say "I missed a few days in the gym." How often do we hear "I didn't get a deep sweat this week"?

Sweat is not a symbol of weakness or impurity; it is a testament to the magnificent resilience of our bodies. It is an expression of our inner strength, a tangible manifestation of our vitality. Sweat carries with it the stories of our physical efforts, the release of accumulated toxins, and the revitalization of our being. It is a natural and essential process that has shaped our evolution as humans, connecting us to our ancient ancestors who also understood the extraordinary power of sweat.

In a world saturated with quick fixes, magic pills, and elusive promises of effortless well-being, we have ignored a pillar of health: sweat. The simple truth that sweat is a profound gift that our bodies offer us - a means of purification, rejuvenation, and renewal. It is a contributing pathway to vibrant health, mental clarity, and emotional well-being. And, in the modern world, it's more important than ever.

I hope, in the same way the world embraced the value of gyms and exercise over the past two centuries, people begin to view heavy sweat-inducing practices in the same light.

It is time to celebrate sweat - to honor its role in our physical, mental, and emotional transformation. Let us create a culture that embraces the critical importance of sweat to our health in the modern world, where saunas and other sweat-inducing practices are not seen as luxuries or indulgences but as essential tools for long, vibrant, healthy lives.

Acknowledgments

Writing this book has been a deeply humbling experience. As I sat down to list every name that influenced its creation, my first thought was "I wrote this book." But, after taking a moment to reflect, I was struck by the vast amount of people I consider to be a part of this book's journey. From a single comment that convinced me of the book's worth to an insightful YouTube video or a motivational quote in a podcast, every interaction played a part.

It's startling to realize that most of these individuals will never know the impact they had. A passing conversation, an encouraging word, a blog post, an interview, a close friend's insight, a Zoom call, early readers, a founder's thoughts on YouTube, or a book I read—all these moments converged to shape this book. It's almost surreal to think that without even one of these influences, this book might not have come to be.

I want to acknowledge some of the many who left their mark on "Sweat":

Family & Friends:

David Glaser, Jordan Glaser, Catherine Ross Glaser, Arnold Ross, Lila Ross, Daniel Glaser, Ellie Glaser, Robin Katz, Norman Leben,

Rami Fattouche, Marshall Borden, Connor Kernochan, Andrew Gross, Pat Walls, Diego Gallovich, Julie Sykora, Taylor Smits, Eve Morey Christiansen, Jason Diamond, Seth Diamond, Alex Kennelley, Luke Walsh, Brooke Whitney, Richard Cardone, Jesse Solomon, Sylvia Goldman, Jacob Topfer, & Geri Topfer.

Interviewees & Collaborators:

Aly Cohen, Meridith Cass, Anne-Marie Principe, Marcia Sabol, Dan Root, Harold Zeliger, Katie Kaps, Andy Kaps, Anoosh Arevshatian, Carly Layne, Owen Sammarone, Lisa Klous, Mohammed Rashik, Mark Schall, Isabella Masucci, Alex Troitzsch, & Ben Keene.

Inspirational People:

Taylor Pearson, David Senra, Shane Tyler Milson, Rudolf Steiner, James Nestor, Alex Hormozi, Meg Jay, Chervin Jafarieh, Justin Mares, Bret Bouer, Tim Ferriss, Jason Silva, Daniel Coyle, Mark Hyman, Dr. David Hawkins, Walter Isaacson, Ron Teeguarden, Dave Asprey, Patrick Bet-David, Richard Koch, Paul Millerd, Ryan Holiday, Tanya Cross, Dr. John Demartini, Donna Perrone, David Perell, Daniel Coyle, Gary Wilson, Jason Polish, Ben Greenfield, & the Starter Story community.

Notes

For a full bibliography with updated and expanded notes, visit sweatthebook.com

Chapter 2: Sweat History

The Native American Sweat Lodge - by Joseph Bruchac

Chapter 4: Toxins: A New Perspective

Body burden study: **https://www.ewg.org/research/body-burden-pollution-newborns**

The Toxin Solution - Dr. Joe Pizzorno

The Autoimmune Epidemic, Dr. Douglas Kerr, M.D., Ph.D.

Lead exposure and development disabilities in preschool-aged children **https://pubmed.ncbi.nlm.nih.gov/28257404/)**

Arsenic methylation capacity and developmental delay in preschool children in Taiwan (https://pubmed.ncbi.nlm.nih.gov/24698386/)

Prenatal exposure to persistent organic pollutants in association with offspring neuropsychological development at 4years of age: The Rhea mother-child cohort, Crete, Greece (https://pubmed.ncbi.nlm.nih.gov/27666324/)

Exposure to lipophilic chemicals as a cause of neurological impairments, neurodevelopmental disorders and neurodegenerative diseases (https://www.ncbi.nlm.nih.gov/pmc/articles/PMC3967436/#CIT0087)

The Body Toxic by Nena Baker

Chapter 5: The Science of Sweat and Detoxification

Stephen Genuis: https://pubmed.ncbi.nlm.nih.gov/21057782/ , https://pubmed.ncbi.nlm.nih.gov/22253637/

Exercise study: https://www.ncbi.nlm.nih.gov/pmc/articles/PMC3312275/

Chapter 7: Niacin Sauna Protocol

Niacin: The Real Story: Learn about the Wonderful Healing Properties of Niacin
   by Abram Hoffer, M.D., Ph.D.

Sauna Detoxification Using Niacin: Following The Recommended Protocol Of Dr. David E. Root by Daniel Root and David Root

Index

About the Author

Justin Glaser graduated from the University of Michigan in 2017 with a B.A. in English (minor in Entrepreneurship). He currently resides in New York City.

For more information, visit www.sweatthebook.com

Printed in Great Britain
by Amazon